LONG ISLAND and the Legacy of EUGENICS

· Station of Intolerance ·

MARK A. TORRES

Published by The History Press
Charleston, SC
www.historypress.com

Cover art courtesy of Gerard DuBois.

First published 2025

Manufactured in the United States

ISBN 9781467158336

Library of Congress Control Number: 2024945299

For my dearest Lola and our children, who complete me.

CONTENTS

AUTHOR'S NOTE

All authors are, or at least should be, creatures of research. This is particularly the case with those who write historical nonfiction. To write competently, we must make painstaking efforts to master (to all extents possible) the subject matter that we are writing about. Since the goal is to create durable and enriching work on important history, there are simply no shortcuts to research.

In 2021, I released my first historical nonfiction book, titled *Long Island Migrant Labor Camps: Dust for Blood*. That was the first book to ever cover the history of the migratory farm labor system on the East End of Long Island, New York. My background as a labor law attorney, a profession that exists for the purpose of helping those in the workplace, provided a never-ending source of inspiration for me to write that book. *Dust for Blood* was a dramatically challenging project mainly because there were no primary sources covering that specific topic. However, since I was writing about an important part of twentieth century history on Long Island that had never been chronicled before, I was more than willing endure the challenge. Contrarily, the important and dramatic history of eugenics, and the horrors it reaped, has been widely studied for many years by notable authors, historians, professors and researchers. As such, there was an abundance of resources available for me to learn this history before writing about it.

Historical topics tend to be quite voluminous, and while many authors provide a broad account of a particular subject, I strongly believe that a narrow and specific focus within an overall historical topic must also be identified. Yet locating such a niche element within a broader history is also challenging. True to my roots as an author of Long Island history, I have always strived to present topics from the oft-neglected local perspective. Thus, this book is not intended to merely serve as a broad retelling of the history of eugenics. Instead, it focuses on investigating the local origins, characters and stratagems employed by the Eugenics Record Office in Cold Spring Harbor, which, for nearly three decades, served as the global headquarters of the eugenics movement. Since this history has never been told from this specific viewpoint, it was a mission that I felt obligated to complete.

Armed with the lens of this local perspective, I approached my research with the same vigor as I had done with *Dust for Blood* and treated the project as if eugenics had never been written about before. My journey led me to study the archival records at numerous facilities, including the Cold Spring Harbor Laboratory and Archives on Long Island; the Rockefeller Archive Center in Sleepy Hollow, New York; the American Philosophical Institute in Philadelphia; Truman State University in Kirksville, Missouri; and the National Museum of Health and Medicine in Silver Springs, Maryland. I also studied several documentaries and other materials in various libraries and in online databases. The information I amassed from these meticulously preserved archives provided sharp insight into the origins, inspiration and machinations of the American eugenics movement, while never losing focus on the fact that it all emanated from a small hamlet on Long Island.

At every facility I visited, I pored through multitudes of boxes, each filled with thick folders containing thousands of items. I combed through countless documents, pedigree charts, letters, court reports, legal statutes, court cases and photographs. I delved into the personal writings of the key drivers of the eugenics movement and read numerous books, reports and articles that they and others had written on the topic. I also reviewed the medical records of many individuals who were studied by eugenicists, while remaining sensitive not to share any personal information out of respect for their privacy. Through it all, I came to understand how eugenics became such an accepted and normalized part of society in the United States and throughout the world during the twentieth century.

Eugenics presented all the trappings of a true science, with the stated purpose that it was meant for the betterment of humanity. Seeing how this developed more than a century after its inception, I came to keenly understand how the manipulation and methodologies formed and took hold of the public at large. I also learned that the key to this success was not limited to mere plotting behind closed doors and in cavernous boardrooms. To the contrary, such planning was followed by a robust public campaign of relentless indoctrination until eugenics became embraced by all facets of American society, including science, medicine, academia, governance, society, popular culture and other popular movements of the time.

Presenting such a complex global history, which spans more than half a century, in a relatively simple and organized manner proved to be another great challenge. After all, eugenicists spent their lives developing and promoting eugenics as a true and complex science needed to save humanity. Thus, an account of this history could not be relegated to merely following the chronological order of events. Instead, my focus was to compartmentalize the many different aspects of this history in several parts and chapters while maintaining a flowing narrative. This, I felt, was the only way to cogently deconstruct the myth of eugenics.

Accordingly, this book is presented in three different, albeit interconnecting, segments. Part I explains the origins and foundations of the American eugenic movement, which is highlighted by the methods utilized by Charles Davenport to establish the Eugenics Record Office in Cold Spring Harbor. This part also details the many aspects of U.S. society that eugenics infiltrated. Part II presents the methods and operations of the practice of eugenics. This section highlights the methodologies of gathering the hereditary data that was used to fuel eugenics as a science, as well as the implementation of marriage restrictions and sterilization laws aimed at eliminating the large number of so-called defectives in the nation. This section also details the most dangerous elements of eugenics, including the direct connection between the American eugenicists and their counterparts around the world, including in Nazi Germany. Finally, Part III of this book chronicles the downfall of the Eugenics Record Office and the ultimate discrediting of eugenics as a scientific field. This final section also explores the enduring and cruel legacy of eugenics.

The quest to perfect our species was not a new one. Such an ideal has long been sought by humans. However, the problem with such aspirations lies in these questions: Who decides the standards of perfection? And, more importantly, what is to be done with those who fall below the arbitrarily created standards? The horrors that lie within the answers to these questions proved to be everything that was wrong with eugenics. Even more worrisome is the fact that there are some practices today that appear to be eugenic in nature, including new technology that strives to create "perfect" children. It is my greatest hope that my work, in some way, can help stimulate further interest in and knowledge of this topic lest we be doomed to suffer the consequences of a movement like eugenics again.

ACKNOWLEDGEMENTS

This book is dedicated to my loving wife, Lola, whose spirit, strength, love and belief in me, along with her much-needed critical thinking, careful analysis and diligent research, have made this book possible. It took a surprising amount of courage to leave my comfort zone of labor-related topics and delve into the dark history of a pseudoscience interwoven with complex issues of racism, classism, culture, poverty, intolerance, disease, disability and more. Frankly, my courage and inspiration to write this book would never have surfaced without her persistently urging me to pursue this subject that has had such a profound impact on humanity.

This book is also dedicated to my children, Isabella, Jake and Olivia, who continue to inspire me every day to be a better father, person, teacher and writer. I would also like to thank my loving mother, Grace, whose strength knows no bounds; my late father for always watching over us; and my family, friends and colleagues for their dedicated and unending support.

This book is also dedicated to the thousands of people across the globe who have suffered the adverse consequences, in any form, of eugenics. It is my greatest wish that this book will inspire greater awareness of and learning about this history so that it can never be repeated in any form.

I have a great amount of respect and appreciation for the many individuals whose enthusiastic support also made the completion of this book possible. They include but are by no means limited to Zoe Ames, Bethany J. Antos, Laura E. Cutter, Joseph DiLullo, Karl Grossman, Trenton Streck-Havill, Steven Klipstein, Melanie Cardone-Leathers, Robert Lewis, Mary Longan,

Paul A. Lombardo, Iris Lopez, Annie Moots, Jessica Gonzalez-Rojas, Jill Santiago, Stephanie Satalino, Johanna Schoen, J. Banks Smither, John Strong, Christopher Verga and Sandi Brewster-Walker. The following organizations, and the wonderful people associated with them, provided tremendous support for my research: the American Philosophical Society, the Cold Spring Harbor Laboratory Library and Archives, The History Press, the National Museum of Health and Medicine, the Rockefeller Archive Center and the Special Collections and Museums at Truman State University.

INTRODUCTION

On November 21, 1946, the trial of *United States of America v. Karl Brandt, et al.* took place at the Palace of Justice in Nuremberg, Germany. Brandt, the lead defendant, had served as Reich Commissioner for Health and Sanitation. He was also the personal physician of Adolf Hitler. He, and twenty-two other doctors and administrators of the Nazi regime, faced prosecution in a case that became known as the Doctors' Trial.[1] These men were charged with crimes against humanity for practicing barbaric medical procedures on human subjects that included freezing, gassing, forced seawater consumption and a euthanasia program that ultimately killed nearly three hundred thousand people during the Second World War.

To justify these acts of torture, Karl Brandt cited a medical report that called for the compulsory sterilization of fifteen million "mentally deficient" people. He also openly read excerpts from an influential legal case that allowed a state to sterilize a young woman against her will. During his testimony, Brandt stated, "In the same way as the state demands the death of its best soldiers, it is titled to order the death of the condemned in its battle against epidemics and disease."[2] The members of the Nuremberg court ultimately rejected the arguments presented by the defense, and in April 1947, Karl Brandt and six others were convicted, sentenced to death and executed.

Astonishingly, the information that Brandt and his cohorts so desperately relied on for their defense was not derived from Nazi propaganda. Instead, their sources came directly from a report published in 1914 by the Eugenics Record Office in Cold Spring Harbor, New York. And the legal precedent they cited in defense of their crimes was from a 1927 United States Supreme Court case that upheld a request from the State of Virginia to forcefully sterilize a young woman against her will because she was arbitrarily deemed to be "feebleminded."

What connection did an administrative office four thousand miles away in a small town on Long Island have with the Nazi regime that plotted and carried out the systematic torture and murder of millions of human beings based on race and disability? How could the doctors who committed such heinous acts have drawn inspiration and support from American jurisprudence as a critical element of their defense? The connection was eugenics: the pseudoscience that dominated much of the twentieth century and was premised on the racist, classist and misguided belief that mental, physical, developmental and behavioral traits of human beings were all inheritable and must be eliminated to save the human race.

Although it was promoted as cutting-edge science, eugenics was a social philosophy that aimed to develop a master race of human beings with the purest blood and the most desirable hereditary traits. This so-called science had two fundamental factions. The first, often referred to as "positive eugenics," was premised on the theoretical belief that the faulty traits that plagued human society could be improved through the selective breeding of people with superior traits. The idea was that if humans could breed better, then our species would be better overall. The other, and far darker, component was referred to as "negative eugenics," which aimed to discourage or outright prevent the reproduction of people who were declared genetically unfit.[3] Negative eugenics was driven by the premise that society would dramatically improve if the millions of Americans who were deemed mentally, physically or morally undesirable were "eliminated from the human stock" by means of segregation, sterilization and even euthanasia. This included the "feebleminded," paupers, criminals, epileptics, the insane, the deformed, the congenitally weak, the blind and the deaf.[4] While human heredity would not begin to be understood by scientists until the 1960s, the social prejudice and practice of eugenics dominated scientific objectivity for more than half a century.

The desire to perfect the human race is not new. The idea that selective breeding will bear the fittest humans is found in ancient Sanskrit and Greek texts, demonstrating that humans have long aspired to create a utopian society where everyone can flourish. Eugenicists believed that breeding a superior race of people would bring about a universally perfect kingdom on earth. The trouble with these ideals lies in these questions: Who decides the favorable criteria within each society? And, more importantly, what shall be done with those who do not meet those criteria? As historian Nathanial Comfort noted, "The problem with utopias is that they establish a set of aspirations that then blind you to a certain set of consequences. And that can be dangerous."[5]

The legacy of eugenics is undeniably cruel and enduring. In the United States alone, more than sixty thousand forced sterilizations were carried out in more than half the states—and likely many more that were never recorded. A multitude of people throughout the country were classified as undesirable and confined to psychiatric centers during their childbearing years. A bevy of marriage restriction and eugenic sterilization laws were enacted for the purpose of preventing the procreation of the unfit. Eugenically driven immigration laws barring the entry of immigrants from many countries into the United States endured for decades. Globally, eugenics thrived in countries like Argentina, Canada, China, Japan and Norway, and Nazi Germany used it to commit unimaginable atrocities. In some ways, the ideals of eugenics persist today.

Despite its global appeal, eugenics was truly made in America, and the epicenter for this movement was not found in some laboratory or governmental facility. Instead, the science was developed at the Eugenics Record Office (widely known as the ERO). This small administrative building was located in Cold Spring Harbor, Long Island, approximately forty miles from New York City. Under the direction of one man, a biologist named Charles Davenport, the ERO operated from 1910 to 1939, where it was used to house massive amounts of hereditary records collected from around the country. The information stored at the ERO not only provided the necessary data for the study of eugenics; it was also used to amplify the colossal movement that was to follow. At the time, scientists had not yet solved the genomic code, and a true study of this science would require generations of human lifespans. Driven by the extreme and immediate desire to rid humanity of degenerative traits, eugenicists short-circuited the scientific process and replaced it with unfounded assumptions, gleaned from the data amassed by eugenic field agents who were dispatched across the country, to reach biased and predetermined conclusions that were driven by racist and classist beliefs. Newly adapted intelligence tests were administered en masse to falsely link people who exhibited low intelligence and antisocial behavior with degenerative traits. These people were then diagnosed using medical terms such as *feebleminded*, *moron* or *imbecile* and subjected to harsh eugenic measures.

The ERO trained and dispatched numerous assistants, many of whom were female college graduates, across the country to interview families, searching for so-called hereditary defects, before returning with data for study and meticulous storage. On Long Island and throughout the nation, eugenicists were granted unfettered access to systematically hunt down and study subjects in prisons, psychiatric centers, Native American reservations, poor rural areas, circus shows at Coney Island and other locations, all in an effort to further legitimize the science. Volumes of reports were produced and distributed. Committees were established to discuss a growing series of eugenic measures, some of which even included mass euthanasia. Elaborate conferences and events were held, drawing dignitaries from all over the world. Scientists collaborated with ERO officials to hone this burgeoning movement. Eugenics was practiced widely throughout the country, and anyone who dared to challenge it was publicly denounced. In short, the ERO perfectly packaged eugenics in the necessary trappings of scientific, academic and administrative fields and touted it as the savior of humanity.

Buoyed by a steep and continuous influx of funding from wealthy, progressive-minded donors and support from most of the leading intellectuals of the time and glamorized in popular culture and mass media, eugenics grew so popular that it eventually became woven into the fabric of American society. Lectures were held and many books written on the topic. Eugenics was taught in hundreds of the most renowned universities in the country and glorified with robust advertising campaigns, cinematic productions and public contests seeking the "fittest" of families.

State legislatures codified laws that led to dramatic increases in institutionalization and the forced sterilization of thousands of unfortunate victims. When these laws were subjected to legal scrutiny, ERO officials had a direct hand in ensuring that they would be upheld. In 1927, the United States Supreme Court, in the infamous *Buck v. Bell* case, ruled that the State of Virginia could sterilize a young woman named Carrie Buck against her will. In the near unanimous decision that has never been overturned, Justice Oliver Wendell Holmes Jr. wrote:

> *It is better for all the world, if instead of waiting to execute the degenerate offspring for crime, or to let them starve for their imbecility, society can prevent those who are manifestly unfit from continuing their kind.... Three generations of imbeciles are enough.*[6]

Fueled by eugenics propaganda and overt racism, lawmakers in Congress passed laws restricting the entry of immigrants into the United States to as little as 3 percent. The impact of the American eugenics movement was profound: it directly inspired the murderous Nazi regime to commit atrocities during the Second World War under the banner of this false science.

How did the Eugenics Record Office function? Who directed such an elaborate operation? Who funded it and why? Most importantly, how did this small, unassuming facility have such a profound impact on world history? *Long Island and the Legacy of Eugenics: Station of Intolerance* presents an unflinching study of the local origins and operations of the American eugenics movement that spread throughout the world.

PART I

FOUNDATIONS

Chapter 1

PLANTING THE SEEDS

All imperfection must disappear.
—Herbert Spencer, Social Statics, *1851*

How do we understand eugenics? The name alone suggests a complex science far too difficult for the layperson to even begin to understand. Over the years, the topic was studied, written and lectured about and even propagandized until it became a part of the fabric of American society. As a result, eugenics embodied all the trappings of an accepted and verifiable science, and for eugenicists, this was the point. In short, eugenics is defined as the pseudoscience that strives for the improvement of the hereditary qualities of a race or breed.[7] Initially named by Sir Francis Galton, the cousin of Charles Darwin, the theories of eugenics began in the nineteenth century, but eugenics really took flight in the early twentieth century. In 1911, a leading American eugenicist named Charles Benedict Davenport wrote:

> *Eugenics is the science of the improvement of the human race by better breeding or, as the late Sir Francis Galton expressed it—"The science which deals with all influences that improve the unborn qualities of a race." The eugenical view is similar to the agriculturalist who, while recognizing the value of culture, believes that permanent advance is to be made only by securing the best "blood."*[8]

Eugenicists like Davenport spent most of their lives driven by the deep-rooted belief that certain traits were in fact inheritable and undesirable traits that manifested in crime, disease, poverty and degeneracy were hindering the pure race that the United States was destined to achieve—and must be eliminated. However, this belief ignored the wide variety of socioeconomic factors that often led to these social ills. The eugenic approach was aimed at permanently improving humanity—like an agriculturalist or horse breeder—by securing only the best "blood," and the best method to accomplish this was believed to be the marriage of only those with the best traits, who would then produce offspring that would possess superior physical and mental traits. As Davenport stated, "Marriage is an experiment in breeding; and the children, in their varied combinations of character, give the result of the experiment."[9]

Despite being presented as a legitimate science, eugenics was truly a social philosophy driven by racist and classist ideals that were used to brand and condemn the downtrodden, who were declared "feebleminded, insane, epileptic, inebriate, criminalistic and other degenerate persons."[10] Terms like *better breeding*, *germ-plasm* and *pure blood* became scientific slogans for a eugenics movement that was well funded by wealthy, progressive-minded Americans, legitimized by the most prominent medical professionals and elite academic thinkers of the time, relied on by government legislators to enact strict immigration laws and even reinforced by the United States Supreme Court.

Eugenics was not some fringe practice that rose from obscurity. Instead, it was spawned amid a confluence of societal factors that gave rise to a negative fervor primarily centered on the poor, the downtrodden, the diseased and the mentally deficient, who collectively came to be viewed as a social plague. To understand the origins of eugenics, we must first look across the Atlantic Ocean at how the English dealt with those in their society who were less fortunate.

By the early sixteenth century, a series of laws had been enacted in England aimed at dealing with the impoverished population.[11] Two categories were created, the first of which was the so-called deserved poor. This group included the youngest and oldest of citizens, along with the infirm and those whose families faced insurmountable financial difficulties. The other category was deemed the "undeserved poor," which included criminals such as highwaymen, pickpockets and beggars. This latter group was deemed the worst of society, and the law inflicted harsh punishments on them. Most notably, poverty, which was also known as vagrancy, was also criminalized.

Although tending to the poor was an expensive endeavor, English leaders understood that inaction would lead to social unrest. Thus, in 1572, compulsory poor law taxes were established, obliging each community to fund poorhouses and other facilities to provide care for their diseased and deranged. As a result, the poor and helpless became isolated from the rest of society. Over the next three centuries, as the populations of slums and poorhouses swelled, England was forced to amend its laws and policies on the treatment of the poor. The costs of funding these institutions were growing, and the ruling class began to rebel against the "taxing of the industrious to support the indolent."[12] Over time, the greater public began to view the impoverished as subhuman, and the added costs of caring for the poor in the country's many publicly funded poorhouses, asylums and orphanages came to be viewed as a social plague.

In 1798, an English economist named Thomas Malthus published a book titled *An Essay on the Principles of Population*.[13] In what ultimately became known as the Malthusian theory, Thomas Malthus reasoned that since our planet has limited resources, humans must adopt strict population control measures. He also argued that charitable efforts to assist the poor promoted further poverty over the generations. Malthus also complained about the unjust economic structures that perpetuated the problem of poverty. However, those who championed the Malthusian theory selectively focused on what he saw as the negative consequences of caring for the poor.

In 1851, an English philosopher, biologist, anthropologist and sociologist named Herbert Spencer published a book titled *Social Statics*. Spencer championed the belief that man must follow the laws of pure science and not the will of God. In fact, and consistent with his philosophy, it was Spencer who first popularized the phrase *survival of the fittest*, and he believed that misery and starvation of the poor was indeed an inevitable element of the laws of nature. He wrote, "The whole effort of nature is to get rid of such, and to make room for the better....If they are not sufficiently complete to live, they die, and it is best if they should die."[14] Less than a decade later, Charles Darwin popularized the phrase *survival of the fittest* in his seminal work *The Origin of Species*. Darwin spoke about the survival process of "natural selection." Ultimately, many of the leading thinkers of the time began to embrace the ideas of Malthus, Spencer and Darwin in a collective philosophy that became known as social Darwinism.

In the late nineteenth century, human genetics was barely understood. One of the early and influential persons with an interest in the subject was Gregor Johann Mendel, an Austrian monk and ardent student of physics, chemistry, mathematics, botany and zoology.[15] Mendel entered the Augustinian monastery in Brünn, Austria, in 1843. With a great interest in agricultural improvement, he maintained an experimental garden and began to research hybridization with different varieties of peas. In a span of a decade, Mendel bred approximately thirty thousand plants and studied the distribution of certain characteristics they possessed, including their height and texture of their seeds. In 1865, he announced to the Natural Sciences of Society in Brünn that the characteristics he studied were determined by hereditarily transmitted elements.

The process of transmission in the pea plants that he studied became known as Mendel's laws of segregation and independent assortment. He relied on the theory that there were two elements for every characteristic. Based on his research, Mendel suggested that the frequency of occurrence of the hereditary elements of the offspring of two plants could be predicted in the same way as predicting the distribution of two different-colored marbles taken from two different bags, with each bag containing a known portion of each color. Mendel spent years studying pea plants trying to ascertain their hereditary traits. After cross-fertilizing more than ten thousand pea plants, he observed that strong or "dominant" traits and weak or "recessive" traits could be predicted. For instance, when he bred a tall pea plant with a short one, it would result in a tall pea plant. Thus, he believed that the trait for tallness was dominant over the short trait, and since tallness was a trait that was preferred by agriculturalists, it was more desirable to produce the tall plants instead of the short plants. Conversely, when he bred pea plants with wrinkled skin and pea plants with smooth skin, he observed that the resulting plants usually had wrinkled skin. Thus, he theorized that the wrinkled skin trait dominated over the smooth skin trait. Since agriculturalists deemed the wrinkled-skin traits less desirable, Mendel believed that the wrinkled skin trait was "corrupting" the more desirable smooth skin trait.[16]

In 1866, Mendel published his findings in the journal *Proceedings of the Brünn Natural Sciences Society*, but his work went generally unnoticed for many years. It remains unclear why Mendel's work remained dormant for so long, but it is believed that it went largely forgotten in favor of the more popular work of Charles Darwin and his theory of evolution and natural selection, which focused on the adaptation of species, while Mendel's work focused on the transmission of characteristics for stability of species. However, after Mendel's death in 1884, biologists began to revisit his work, and this rediscovery of so-called Mendelism would serve as a foundation for eugenics.

At around the same time as Mendel was performing his work, Sir Francis Galton began to be recognized for his own scientific achievements. In 1822, Galton was born into a wealthy Birmingham family who amassed great wealth from gun manufacturing and banking.[17] He was the youngest of seven children and a cousin of Charles Darwin. In his youth, he was looked after by his loving and well-learned sister Adele, who was twelve years older than him and bedridden because of curvature of the spine. She managed to teach Galton many lessons before he went off to school. At just over two years old, Galton could read; at four he could write and do arithmetic; and at eight he comfortably understood classical Latin texts.

Galton was raised as an Anglican, and this enabled him to attend England's leading universities, which at the time were restricted to members of the Church of England. Galton began his studies at King's College Medical School, but he detested the study of medicine, and in 1840, he enrolled in Cambridge University to study mathematics and became well educated in the field of statistics. He attempted to complete his studies in just three years before suffering a nervous breakdown, which forced him to take some time off from his studies before earning his degree. After the death of his father in 1845, Galton traveled throughout southern Africa. He then met and married Louisa Butler, and the couple settled in a home near Hyde Park.

Galton always distinguished himself by his keen ability to recognize patterns in most situations. He lived by a general philosophy: "Wherever you can, count."[18] At public events, he would observe the number of people who fidgeted in the audience. In private, he would study the patterns of waves in his bathtub. Galton's work eventually yielded important scientific results. In 1861, he distributed a questionnaire to the weather stations of Europe seeking information about the weather conditions for the month of December and developed a pattern that led to the formation of the world's first weather maps. In 1863, he published *Meteorographica, or Methods of Mapping the Weather*, which greatly advanced the science of meteorology. In response to a rash of unsolved crimes in England, including the infamous Jack the Ripper case, Galton studied the patterns of human fingertips and discovered that the raised ridges on the fingertips were unique to each individual. He released a book on his findings, and the use of fingerprints revolutionized the field of criminal forensics.

Galton also studied the patterns of various qualities of human beings and authored a book published in 1869 titled *Heredity Genius*. He believed that not only physical qualities such as hair color and height could be transmitted but mental, emotional and creative qualities as well. He also thought that people's abilities and features were achieved by no accident and could even be calculated and honed into a "highly gifted race of men by judicious marriages during several consecutive generations."[19] Galton

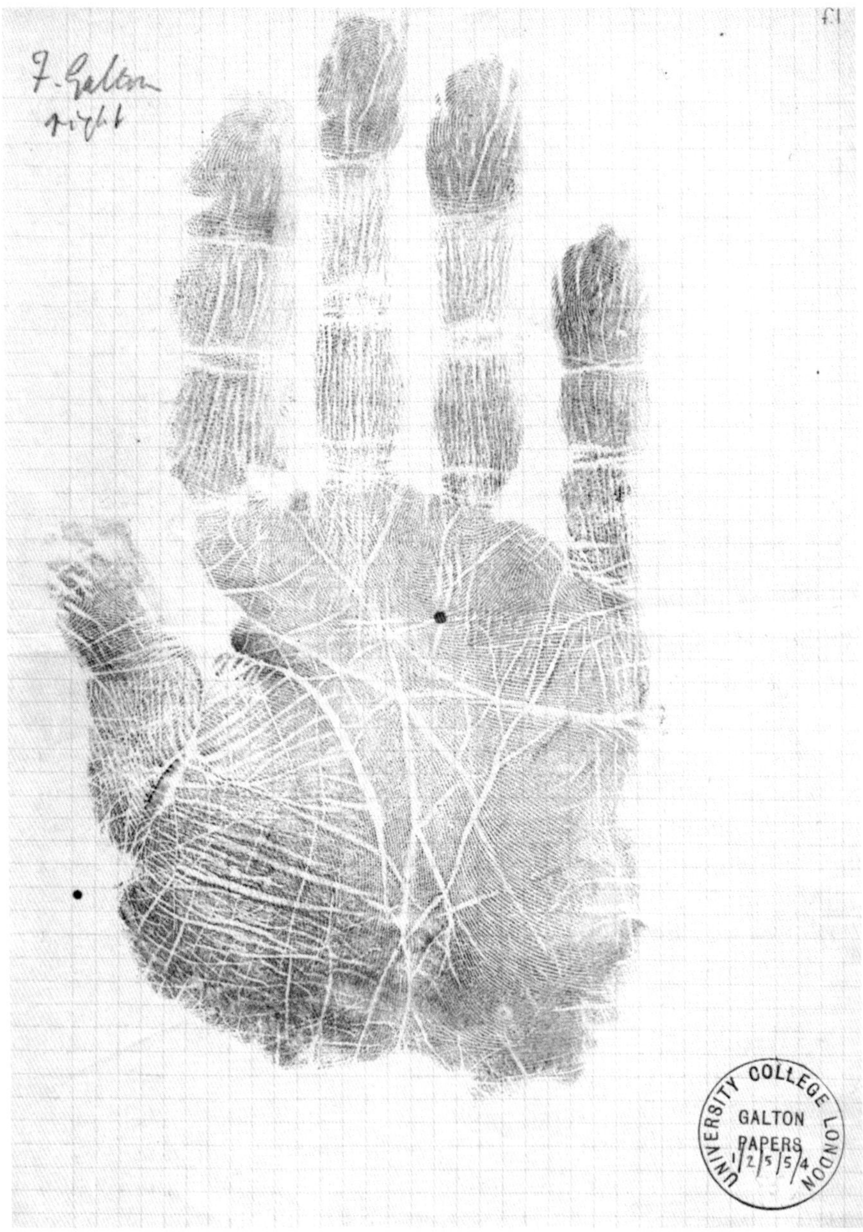

Above: Handprint of Francis Galton. *Photo courtesy of UCL Special Collections Library, Culture, Collections & Open Science, London, England.*

Opposite: Fingerprints of Francis Galton with written notes. Galton's study on human fingertips in the late nineteenth century revolutionized the field of criminal forensics. *Photo courtesy of UCL Special Collections Library, Culture, Collections & Open Science, London, England.*

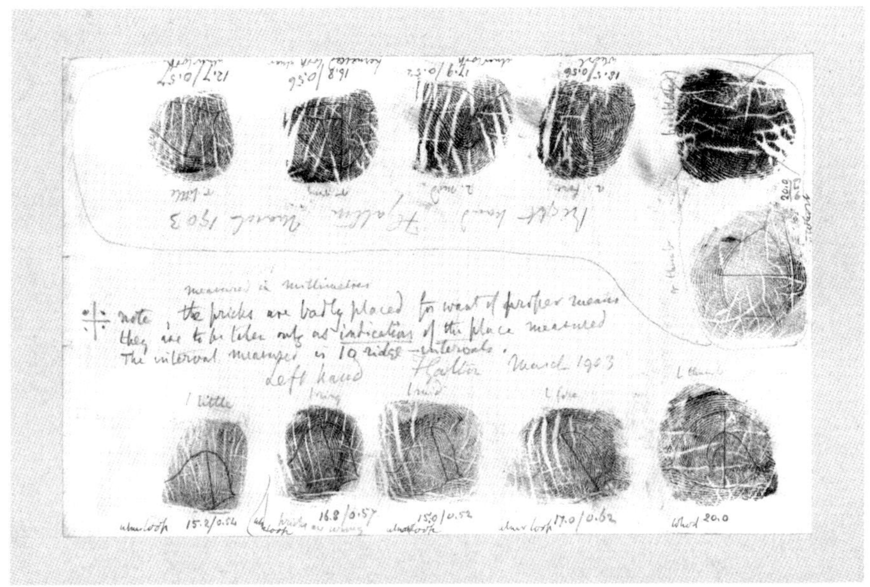

suggested that the breeding of the best people would evolve mankind into a super species, and he sought to create a regulated marriage process whereby members of only the finest families would marry carefully selected spouses. He would often ask, "Could not the undesirables be got rid of, and the desirables multiplied?" Although he was unable to identify a predictable or numerical formula for such an ideal, Galton was convinced that the secret of scientific breeding would eventually be revealed, and it would change mankind forever. In 1883, Galton published *Inquiries into Human Fertility and Its Development*. After coining several names for his so-called new science, he jotted various Greek and English letters on a piece of paper and eventually linked the words *born* and *well* before settling on the word *eugenics*.[20]

The founding fathers of eugenics in England had formulated the theoretical concepts of human hereditary research. It was only a matter of time before it caught on in the United States, and of the many individuals and groups who helped establish and promote eugenics from theory to practice, none was more influential than an American biologist named Charles Davenport who was directly responsible for the establishment and operation of the Eugenics Record Office, which for more than three decades would serve as the eugenics capital of the world. Davenport also led the movement that would ultimately springboard eugenics into a global phenomenon.

Chapter 2

THE ARCHITECT (DAVENPORT'S CRUSADE)

I may say that the principles of heredity
are the same in man and hogs and sunflowers.
—Charles Davenport, 1912

Charles Benedict Davenport was born in 1866 at his family's summer home, affectionately known as Davenport Ridge, near Stamford, Connecticut.[21] He was the eighth child of Amzi Benedict Davenport and Jane Joralemon Dimon Davenport. His father was a schoolteacher who also operated a successful real estate business in Brooklyn. His mother was the daughter of John Dimon, a successful carpenter and builder.

The Davenport family spent winters in their home in Brooklyn Heights. During his youth, Charles Davenport was homeschooled and spent time working in his father's office. On November 26, 1879, at the age of thirteen, he enrolled in Brooklyn Collegiate and Polytechnic Institute. In 1886, he graduated with a bachelor's degree in civil engineering and briefly worked as a surveyor for the Duluth, South Shore and Atlantic Railway before enrolling at Harvard University in September 1887.

At Harvard, Davenport excelled in the field of zoology. In 1889, he earned an undergraduate degree and began working at a biological station at Woods Hole, Massachusetts, while continuing his studies. In 1892, Davenport earned a PhD and began a career as an instructor at Harvard, where he taught introductory zoology courses until 1899. During that time, he met

Undated photo of Charles Davenport standing outside. *Photo courtesy of the Truman State University, Pickler Memorial Library, Special Collections and Museums.*

Gertrude Crotty, the daughter of William and Millia Crotty of Burlington, Kansas. Like Davenport, Crotty was a student before becoming an instructor at Harvard, and they both studied marine biology in Newport, Rhode Island, in the summer of 1893. The couple married a year later and welcomed their first child, Millia Crotty Davenport, on March 30, 1895. Their second daughter, named Jane Joralemon Davenport, was born on September 11, 1897. Later, the couple had a third child, Charles Benedict Davenport Jr., born on January 11, 1911, but after contracting polio, he died on September 5, 1916.

In the spring of 1898, Charles Davenport was appointed director of the summer school program at the Brooklyn Institute of Arts and Sciences Biological Laboratory at Cold Spring Harbor, New York.[22] The following year, he became an associate professor at the University of Chicago. In November 1900, Davenport became one of the first scientists in the United States to write about the rediscovery of Gregor Mendel's laws of inheritance. In 1902, the Carnegie Institute of Washington was founded, and Davenport immediately began to lobby the group to invest in the establishment of a center for genetics at Cold Spring Harbor. It may not have been known at the time, but the forces were beginning to align for the formation of the American eugenics movement, and Charles Davenport would be at the center of it all.

CHARLES DAVENPORT WORKED TIRELESSLY in pursuit of his eugenic goals. He was well educated and fiercely determined. Drawing influence from Gregor Mendel's work on the heredity of peas and Francis Galton's statistical mind, he developed a plan to collect hereditary information from a multitude of families in order to prove that evolution worked in human beings the way it worked in animals and plants. It was to be a revolutionary project, requiring

The Davenport family: Charles Davenport is standing between the couple's daughters, Millia and Jane. Seated is Gertrude Crotty, and on her lap is their son Charles Jr. who, after contracting polio at the age of five, died on September 5, 1916. *Photo courtesy of the Truman State University, Pickler Memorial Library, Special Collections and Museums.*

The Davenport residence in Cold Spring Harbor, New York. *Photo courtesy of the Truman State University, Pickler Memorial Library, Special Collections and Museums.*

the support that would come in the form of several professional relationships Davenport developed.

In 1902, Davenport embarked on a long-awaited journey to Europe with his wife, Gertrude. Their trip included a highly anticipated meeting in London with Francis Galton himself. At this meeting, the two men discussed their belief in the science of eugenics, and Davenport explained his goal of establishing eugenics research in the United States. The meeting was a success, as Davenport was furnished with a letter expressing strong support from the renowned Francis Galton. Armed with the blessing of one of the founding fathers of eugenics, Davenport returned to Long Island with "renewed courage for the fight of quantitative study of Evolution."[23]

A second pivotal relationship for the founding of the American eugenics program existed between Charles Davenport and an American physiologist

Left: Charles Benedict Davenport and Gertrude C. Davenport. *Photo courtesy of the American Philosophical Society.*

Opposite: A eugenics-based propaganda illustration, titled "Steps in Mental Development," purporting to depict various mental levels of "feeblemindedness" in ascending order. Today, terms like *idiot* and *moron* are generally used as insults, but in the early twentieth century, eugenicists propagandized these terms as medical diagnoses to brand citizens and subject them to harsh eugenic measures. *Photo courtesy of the Library of Congress, 1915.*

named Henry Goddard. Henry Herbert Goddard was born in 1866 in Vassalboro, Maine, to Henry Clay and Sarah Goddard, both Quakers.[24] Goddard was an avid student who earned a bachelor's degree in 1887 from Haverford College in Pennsylvania. He spent a year teaching Latin, history and botany at the University of Southern California before returning to Haverford College to earn a master's degree in mathematics in 1889. Later that year, Goddard began teaching in Damascus, Ohio, where he met Emma Florence Robbins; the couple wed soon thereafter.

Goddard later studied psychology at Clark University in Worcester, Massachusetts, and became involved in the movement to help develop laws for the education of children. In 1901, Goddard met Edward Ransome

Johnstone, principal and director of the New Jersey Training School for Feeble-Minded Children at Vineland. They shared an interest in the study of mental deficiencies. In 1906, Johnstone hired Goddard to be director of research on mental deficiencies at the Vineland school.

In 1908, Goddard traveled to Europe to conduct further research on child intelligence. There, he learned about an intelligence test administered to children by French psychologist Alfred Binet and physician Théodore Simon. The purpose of the test was to determine the mental age of children based on their ability to perform certain tasks, like identifying objects or copying patterns. Children were administered the Binet and Simon intelligence test, and if they scored below average for their age group, they would be declared deficient. Goddard translated the test into English and began to administer it to the patients at the Vineland school. He also tested students in the New York City public school system, where, based on the results, he estimated that as many as fifteen thousand students were mentally deficient. Over time, Goddard's version of the Binet and Simon intelligence test became a popular tool used to measure intelligence throughout the United States.

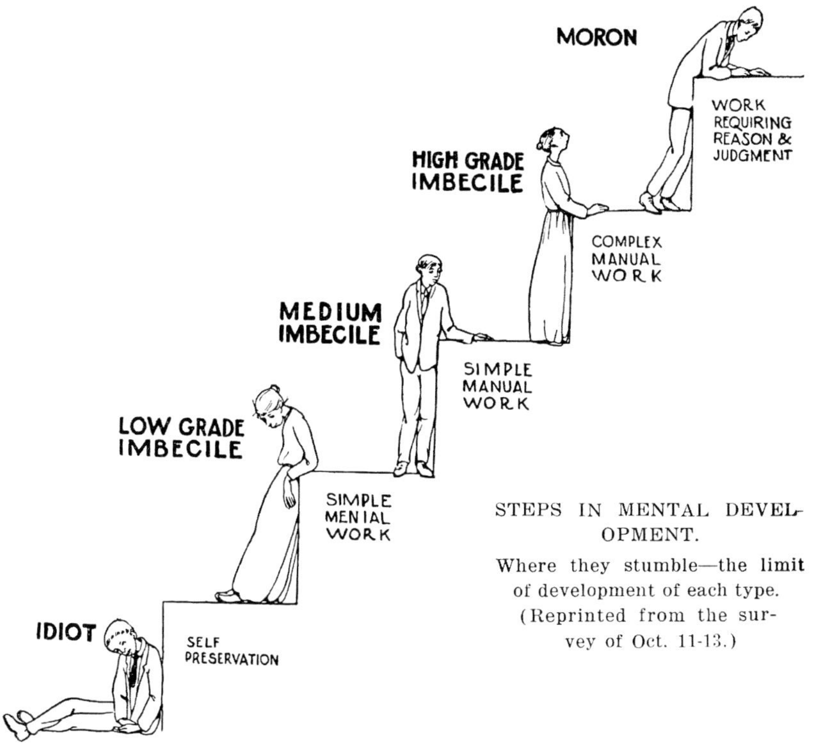

STEPS IN MENTAL DEVELOPMENT.
Where they stumble—the limit of development of each type. (Reprinted from the survey of Oct. 11-13.)

Through his work, Goddard developed terms like *idiot* and *imbecile* as medical diagnoses, but none was more prevalent than *moron* or *feebleminded* in his work. He believed that these were slightly higher-functioning patients who were trapped in a type of evolutionary phase without moral judgment. In 1912, Goddard stated, "The idiot is not our greatest problem. He is indeed loathsome….Nevertheless, he lives his life and is done. He does not continue the race with a line of children like himself….It is the moron type that makes for us our great problem." Later that year, Goddard authored *The Kallikak Family: A Study in the Heredity of Feeble-mindedness*. The best-selling book further popularized his view that feeblemindedness was an inheritable trait that directly contributes to crime, poverty and other problems within society.

The relationship between Charles Davenport and Henry Goddard cannot be understated. It was through their collective beliefs that *mental deficiency* and *immorality* became interchangeable. As a result, eugenics fully embraced the dangerous principle that moral depravity was an inheritable trait. Thus, peopled labeled with terms like *feebleminded* and *moron* became true targets for eugenicists, and such labels served as catch-all diagnoses used to declare and condemn many people as unfit for society. Moreover, Goddard's intelligence test became an important tool for eugenicists all around the world to diagnose so-called defective subjects and further legitimize eugenics as a reliable and established science.

At this time, Charles Davenport was bestowed with an endorsement from Francis Galton for his eugenic work. His relationship with Henry Goddard helped him hone the science by adding catch-all diagnoses and an intelligence test to confirm those diagnoses. Now, he required a trusted assistant who could offer a direct and hands-on approach to the day-to-day operations of his eugenic program. That assistant would be Harry Laughlin. Although Laughlin was hired as the superintendent of the Eugenics Record Office, his involvement in the eugenic program was arguably the most pivotal. In a near evangelical fashion, he served as the political and strategic mouthpiece of Davenport's eugenics movement.

Harry Laughlin. *Photo courtesy of the Truman State University, Pickler Memorial Library, Special Collections and Museums.*

Harry Hamilton Laughlin was born in 1880, the eighth of ten children, to George Hamilton Laughlin and Deborah Jane Ross Laughlin in

Oskaloosa, Iowa. His father was the president of and a teacher at Oskaloosa College and a local minister for the Disciples of Christ.[25] The Laughlin family later moved to Kirksville, Missouri, and George became an English professor at a school that later became known as Truman State University, teaching there until his death in 1895.

While four of his brothers studied to become osteopathic doctors, Laughlin focused his studies on various fields of science. He became the principal of a local high school in Canterville, Iowa, and later married Pansy Bowen on September 12, 1902. The couple never bore children. In 1907, Laughlin began teaching agriculture and nature studies at the Kirksville State Normal School. He later earned a master's degree in biology at Princeton University in New Jersey and conducted his doctoral work in 1917 on mitotic stages in the division of onion root-tip cells. Despite his academic accolades, Laughlin developed a reputation as one who took shortcuts in experimental research. This was known by Davenport, who nevertheless appreciated Laughlin's work and invited him to serve as the superintendent of the Eugenics Record Office. Laughlin would ultimately serve in this role from 1910 to 1921 before becoming assistant director of the ERO from 1921 until its closing in 1939.

Harry Laughlin worked tirelessly to amplify eugenics from the local facility at Cold Spring Harbor to the rest of the world. He served two very important and primary roles in the eugenics program. One directly led to the passage of strict immigration laws. The other was to develop a blueprint for the compulsory sterilization of U.S. citizens who were deemed unfit. This included the creation of a model sterilization law that U.S. states would use to craft their own statutes that could withstand constitutional scrutiny. Later, Davenport and Laughlin maintained strong and dangerous ties with eugenicists in Nazi Germany.[26]

Charles Davenport keenly understood that lucrative funding was required in order to develop his eugenic program. He first turned to a local widow named Mary Williamson Averell Harriman for assistance. Mary Harriman was born in Rochester, New York, on July 22, 1851. Her father was W.J. Averell, a banker and railroad official.[27] On September 10, 1879, she married Edward Henry Harriman, a wealthy railroad investor and builder. The couple had six children. After her husband's death in 1909, Mary Harriman was named the sole executor of his estate, valued at more than $100 million, approximately $3.7 billion in today's money. The widow shared an interest in her late husband's business and actively played an important part in many civic, philanthropic, educational and artistic endeavors, including making a donation of ten thousand acres of land to the State of New York to establish

Left: Mary Williamson Averell Harriman, June 5, 1915. At the time, she was publicly known as Mrs. E.H. Harriman for her marriage to Edward Henry Harriman, a wealthy railroad investor and builder. On her husband's death in 1909, she was named the sole executor of his estate, valued at more than $100 million, approximately $3.7 billion in today's money. She was a fervent supporter of Charles Davenport, and for much of her life, she provided lucrative donations to the American eugenics program. *Photo courtesy of the Library of Congress.*

Below: Mary Harriman in Tuxedo, New York, circa 1915. *Photo courtesy of the Library of Congress.*

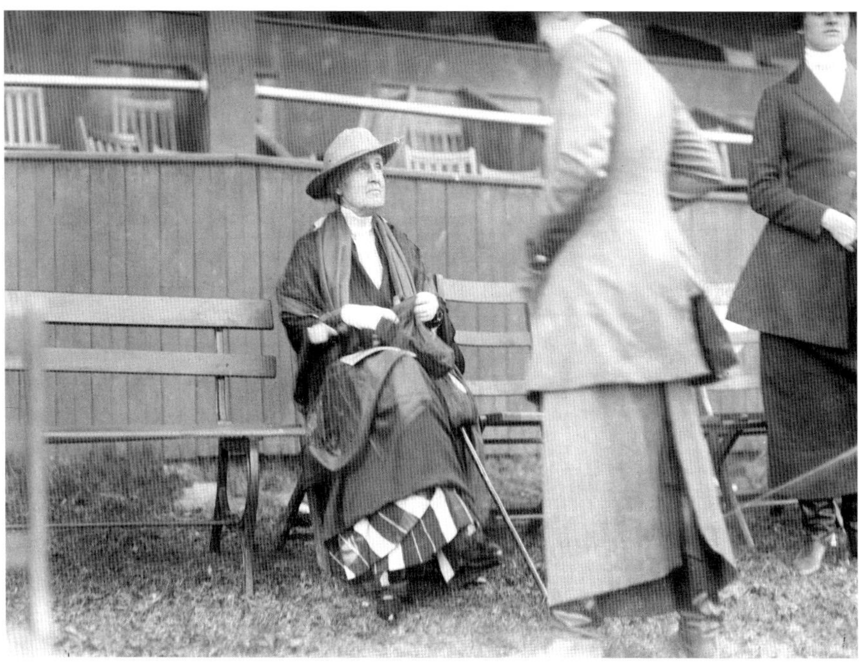

Harriman State Park. Mary Harriman continued with this work until her death on November 7, 1932, at the age of eighty-one.

In order to solicit support from Mary Harriman, Davenport first approached her daughter Mary Harriman Rumsey. Mary Rumsey spent part of the summer of 1905 as a student at the Cold Spring Harbor Laboratory as part of her undergraduate studies at Barnard College.[28] She was fully aware of Davenport's work and believed that eugenics had an important role to play in social improvement. Thus, she eagerly agreed to arrange a meeting between Davenport and her mother.

In February 1910, Charles Davenport and Mary Harriman met for lunch in New York City. The stakes could not have been higher. Davenport desperately needed to secure funding, and Harriman was eager to support a cause that she believed in. The full details of what was discussed at this fateful meeting are unknown, but Mary Rumsey explained that her mother shared in the horse-breeding interests of her late husband, who believed that proper breeding had potential for the improvement of humanity. Davenport likely tapped into this interest, and after some discussion, Mary Harriman agreed to support his eugenics research. Davenport would later write in his diary that it was a "red letter day for humanity."[29] Until her death in 1932, Mary Harriman remained an ardent supporter of Charles Davenport and the eugenics program. Between 1910 and 1918, the total amount of money she donated to the eugenics program is estimated at more than $500,000, equivalent to approximately $11 million in today's money.[30]

Charles Davenport undoubtedly proved to be the architect and lead scientist behind the American eugenics program. He furthered eugenical studies and secured relationships that led to much-needed funding for the program. His reputation as a pioneering biologist was buoyed by the early support of Francis Galton, one of the founding fathers of eugenics. He also developed a working relationship with Henry Goddard, and together, they expanded the science of eugenics to include the newly and arbitrarily developed medical classification of "feeblemindedness." This, along with the establishment of new eugenic testing procedures, would lead to the ensnarement of a vast number of so-called defective Americans. His relationship with Harry Laughlin would greatly expand the philosophy, practice and eventual weaponization of eugenics. Finally, Davenport's relationship with Mary Harriman provided the initial and ongoing financial support needed to develop a thriving program at the Eugenics Record Office in Cold Spring Harbor, which lasted for three decades and played a pivotal role in propelling eugenics throughout the United States and the rest of the world.

Chapter 3

COLD SPRING HARBOR

Quite an empire for us, isn't it?
—Charles Davenport to Gertrude Crotty Davenport, January 24, 1904

Cold Spring Harbor is a charming hamlet located on the shore of Oyster Bay, Long Island, approximately fifty miles outside of New York City. The land was originally inhabited by the Matinecock people, who called the area Wawepex, which meant "place of good water."[31] In 1650, English settlers purchased some acreage of the land and named the area Cold Spring. The community was officially named Cold Spring Harbor in 1825 for its freshwater springs running north through the area and into the harbor. With its proximity to New York City, the town was a suitable location for milling factories that produced wool and other cloth products.[32]

Although not particularly long, the harbor is fairly deep, which made it ideal for use during the peak season of the whaling industry. In the mid-1800s, Cold Spring Harbor supported a fleet of whaling vessels and other ships that carried cargo from Long Island to towns along the Atlantic coast and the Caribbean. Over time, the area became one of Long Island's busiest whaling ports. It also became a popular vacation destination for New Yorkers. The area was described by one reporter as "a beautiful place, almost utterly lacking business, combining most charmingly the country and seashore, and for a long time has been the favorite summering resort of the wealthy but not obtrusively aristocratic class of people."[33]

The Brooklyn Institute of Arts and Sciences evolved from the Brooklyn Apprentices' Library, which was formed in 1824.[34] Toward the end of the nineteenth century, the institute embarked on an ambitious period of growth, which eventually led to the creation of the Brooklyn Museum, the Brooklyn Botanical Garden and the Brooklyn Academy of Music. The institute also looked toward Long Island to expand its burgeoning science program.

On July 7, 1890, the Brooklyn Institute of Arts and Sciences opened a laboratory for biological research at Cold Spring Harbor. The locale was an ideal setting for biological research. This facility also offered summer classes to students eager to learn the field of science, with lectures on various biology topics. Originally built in 1887, the one-and-a-half-story wooden building was owned by the New York Fish Commission, which loaned the entire facility to the laboratory through the generosity of Fish Commissioner Eugene Blackford.[35] The head of the building was located near a beach set amid rolling hills and an abundance of forests, glens and small streams providing a wide array of flora and fauna. Four freshwater ponds supplied crystal-clear drinking water to the Cold Spring creek, which fed into underground wells and provided an abundance of life well suited for the study of biology. Just below the laboratory was the beautiful Cold Spring Harbor, divided by a sandy neck into an inner and outer basin.

The laboratory was fitted with state-of-the-art furnishings, apparatus and appliances needed for biological study. Each student was provided an individual table fitted with compound microscopes, tools and glassware and personal aquariums for the collection and study of specimens. Students were afforded the opportunity to venture out on boats, both in the harbor and on the Long Island Sound to explore the coastline. It was an ideal location for the study of a multitude of biological specimens, including algae, fish, mollusks and insects.

In 1898, Charles Davenport was appointed the new director of the biological research program.[36] Along with his administrative duties, both he and his wife, Gertrude Crotty Davenport, also taught classes in microscopy. Other scheduled courses included lectures on anatomy, botany and zoology. The program also offered evening courses taught by esteemed professors George H. Parker from Harvard and Charles L. Bristol from New York University, along with D.S. Judge from the United States Department of Agriculture.[37]

In June 1904, the Carnegie Institute of Washington leased a ten-acre tract of a land in Cold Spring Harbor.[38] Originally established just two years earlier after a lucrative endowment of $10 million by Andrew Carnegie, the

Carnegie Institute was incorporated by an act of Congress "to encourage, in the broadest and most liberal manner, investigation, research and discovery, and the application of knowledge to the improvement of mankind." Eager to expand this mission throughout the country, the Carnegie Institute secured a fifty-year lease at Cold Spring Harbor, which was already a burgeoning hub of scientific research. By this time, Charles Davenport was receiving an annual salary of $3,500 with a promise of a raise to $4,000 the next year. This salary placed him among the highest-paid professors in the nation. The Davenports had previously purchased a home on six acres near the shore of Cold Spring Harbor and later added an additional nineteen-acre tract of land, which was purchased in Gertrude's name. In a letter he later wrote to his wife, Davenport stated, "Quite an empire for us, isn't it?"[39]

The Carnegie Station for Experimental Evolution, simply known as the Station, was a sixty-foot-long, thirty-five-foot-wide brick building erected adjacent to the Brooklyn Institute of Arts and Sciences laboratory. The building stood two and a half stories high and boasted administrative and private rooms for investigators, a photographic room and several large rooms for studying animals in varying conditions of light, moisture and temperature.[40] All the rooms were powered with electricity and running salt water and were well-equipped for laboratory experiments. A large library contained scientific literature on a variety of topics, including heredity and plant and animal breeding. As director of the Station, Davenport enlisted G.H. Shull (botanist), F.E. Lutz (entomologist), Anne M. Lutz (cytologist) and Mabel E. Smallwood (librarian) to assist with the program.

The study of heredity was the focus of experimentation at the Station. This included work with birds, plants and insects, along with larger animals like sheep, goats and cats. Reporters who were allowed access to this new facility were greeted with the songs of canaries echoing throughout the building. There was a great amount of anticipation surrounding the work at the Station and its quest to unlock the laws of inheritance and heredity. Charles Davenport proudly boasted: "The problem of the origin of species has taken quite a new form in the half century since Darwin's epoch-making work appeared. The work of the Station is not therefore to confirm Darwin, but to go much farther than Darwin did."[41]

Although much of the work toward understanding human heredity was still relatively new, Davenport stated, "Although not strictly within the scope of experimental work, the necessity of applying the knowledge of heredity to human affairs has been too evident to permit us to overlook it."[42] He added that approximately five thousand questionnaires had been sent out to residents throughout the nation, seeking information on human hair and eye color. There was little doubt about Davenport's eugenic aspirations. During an interview, he touted an essay he authored titled "Eugenics, the Science of Human Improvement by Better Breeding," in which he wrote:

> *Were our knowledge of heredity more precisely formulated, there is little doubt that many certainly unfit matings would be prevented, and other fit matings, that are avoided through false scruples, would be happily contracted. Governments spend scores of thousands of dollars and establish rigid inspections to prevent the spread of the coitus disease of the horse, but the spirochete parasite that causes the corresponding disease in man and entails endless misery on hundreds of thousands of innocent children may be disseminated by anybody and is being disseminated by scores of hundreds of thousands of persons in this country unchecked, under the protection of personal liberty. Alas! that so little thought is given to the loss of liberty by the infected children. Marriage of persons afflicted with certain diseases is not only unfit, it is a hideous and dastardly crime, and its frequency would justify regulations far more strict than those which now prevail.*

As work at the Station continued in earnest, plans were also being made to expand the eugenic program. Mary Harriman purchased an eighty-acre tract of land at Cold Spring Harbor, and an office building was erected in 1910 to officially establish the Eugenics Record Office. Mary Harriman made it clear that she was eager "to place the Eugenics Record Office on a more permanent basis."[43] Later on, Harriman donated the eighty-acre tract of land and a gift of $300,000 in securities to the Carnegie Institute, which formally accepted the offer at a trustee meeting on November 19, 1917. Davenport wrote to thank her, stating, "What a fire you have kindled! It is going to be a purifying conflagration some day!"[44]

In 1921, the Carnegie Station for Experimental Evolution and the Eugenics Record Office were combined and placed squarely under the control of Charles Davenport.[45] As a part of the merger, the ERO transferred its collection of 51,851 pages of family information and 534,625 index cards on individuals.[46] The facility's data was meticulously maintained, including records of all expenditures. Over time, more wealthy financiers donated to the program, including John D. Rockefeller, who donated $22,000 over four years.[47] Such financial support from both national and local figures provided instant and ongoing credibility to the new program, which drew more interest from donors and intellectuals of the time.

The ERO was not only active in eugenic research. Under Davenport's direction, the facility also participated in many community-based activities for the benefit of the public. In 1911, the administration formed a local bird club, which drew nearly forty area children to the facility to gaze at the many stuffed birds that were held on exhibit as well as the many birds that frequently visited the sprawling grounds.[48] On September 28, 1921, the ERO hosted approximately 150 members of the second International Eugenics Congress. The guests toured the facility and enjoyed a luncheon at the nearby Piping Rock Club in Cold Spring Harbor.[49] On August 15, 1922, Jane Joralemon Davenport, Charles Davenport's daughter, married Dr. Reginald G. Harris at a ceremony in Cold Spring Harbor.[50] These types of activities helped support local businesses in the area and undoubtedly endeared the ERO to many in the community.

Local reporting also kept the public apprised of matters pertaining to the facility, including news of calamitous events. During the evening of June 5, 1912, a fire erupted at the Eugenics Records Office in Cold Spring Harbor.[51] Several barns and other buildings were destroyed in the fire. Two horses perished, and a large amount of farming equipment and scientific dairy apparatus was lost. The cost of the damage was estimated to be $25,000.

ERO officials also engaged in civic and town improvement activities. Since the office was located in the hamlet of Cold Spring Harbor, which was and still is a part of the township of Huntington, these activities concerned vital public projects. In 1929, Harry Laughlin served as president of the Cold Spring Harbor Village Improvement Society.[52] In that role, he championed a series of improvements, including the installation of street and traffic lights in various locations within the village, which were deemed necessary after two auto accident fatalities, and other road construction projects.[53] The group also finalized plans to develop a fifteen-acre park along Main Street, which was to be named Memorial Park. Financing for the projects came from a mix of local government funds and monies donated by town residents.[54] Other projects proposed by the group included the construction of a new firehouse.

Charles Davenport also engaged in local activities. In 1914, he was heavily involved in the planning to establish a tuberculosis hospital in Nassau County. During a meeting on October 5, 1914, at the Nassau County Association in Mineola, Davenport asked for public support and funding for the project. He also organized poll watchers for the public referendum to decide the fate of the project.[55] Even after his retirement in 1934, Davenport remained active in carrying out local duties, serving as an air raid warden for the Nassau

The Animal House was a biological facility at the Carnegie Station of Experimental Evolution at Cold Spring Harbor, New York, used to study a wide array of birds, plants and insects. *Photo courtesy of the Truman State University, Pickler Memorial Library, Special Collections and Museums.*

The Carnegie Station of Experimental Evolution at Cold Spring Harbor, New York. *Photo courtesy of the Truman State University, Pickler Memorial Library, Special Collections and Museums.*

Undated photo of chicken coops and a garden with the first hybrid corn crops grown at the Carnegie Station of Experimental Evolution at Cold Spring Harbor, New York. *Photo courtesy of the Truman State University, Pickler Memorial Library, Special Collections and Museums.*

Harry Laughlin (*left*) and Charles Davenport (*right*) outside the Eugenics Record Office. *Photo courtesy of the Truman State University, Pickler Memorial Library, Special Collections and Museums.*

This page: The newly constructed eugenics research and administration building at the Carnegie Station of Experimental Evolution in Cold Spring Harbor, New York, circa 1904. *Photo courtesy of the Truman State University, Pickler Memorial Library, Special Collections and Museums.*

This page and opposite: The Eugenics Record Office in Cold Spring Harbor, New York, was built in 1910 with financial support from Mary Harriman and operated until its closure in 1939. In 1917, Harriman officially donated the building, an eighty-acre tract of land and $300,000 to the Carnegie Institute to continue eugenics work on a more permanent basis. In 1921, the Carnegie Institute consolidated the operations of the Carnegie Station of Experimental Evolution and the Eugenics Record Office and named Charles Davenport director of all operations. *Photos courtesy of Cold Spring Harbor Laboratory Archives, New York.*

County Defense Council and a lecturer at an elementary school near his home.[56] He also served as president of the Oyster Bay Township Taxpayer's League and oversaw local expenses on town projects, including the purchase of a new incinerator.[57] He also assisted with various mosquito extermination projects in both Nassau and Suffolk Counties.

The civic duties performed by Davenport, Laughlin, and others at the ERO undoubtedly benefited the township in and around Cold Spring Harbor. The facility also offered many educational programs to the public. Whether these acts were used to shroud the true nature of the work that would eventually be performed at the ERO may never be known, but it is certain that the ERO had found a home in the hamlet of Cold Spring Harbor, where it would remain for nearly three decades.

Chapter 4

MADE IN AMERICA

The laws of nature require the obliteration of the unfit and human life is valuable only when it is of use to the community or race.
—*Madison Grant,* The Passing of the Great Race, *1916*

During the first half of the twentieth century, a growing number of scientists began to embrace eugenics as a true hereditary science. Eugenics also benefited from the financial support of progressive-minded donors who were eager to cure the many social ills of the time. These supporters were most often middle- to upper-class citizens, educated, White, Anglo-Saxon and predominantly Protestant Americans.[58] Over time, eugenics was promoted, glamorized and even preached to many other parts of American society, including academics, politics, popular culture, religion and other social movements. As a result, eugenics became a widespread and seemingly unstoppable force that was engrained into the fabric of American society. This chapter explores some of the major societal areas where eugenics took root.

If eugenics was to be accepted as a legitimate science, it was paramount that it be embraced by American academia. This was necessary to not only establish the credibility of this fledgling science but also to ensure its future by promoting it to students who would inevitably become medical practitioners, teachers and leaders of the nation.

During much of the early to mid-twentieth century, eugenics was taught, at least in some form, at the most prestigious academic institutions in the country, including Harvard, Johns Hopkins, Princeton and Yale. Students learned the field of eugenics—often consolidated with courses on biology, genetics, psychology, zoology or even farming and animal husbandry—in hundreds of American colleges. According to an ERO report issued in February 1916, over 254 colleges embraced eugenics as part of their curriculum.

At Boston University, eugenics was taught to students at the School of Theology, and a course at the Massachusetts Institute of Technology (MIT) presented "experimental advances in the study of biometrics, heredity and eugenics." The University of Oregon offered courses in both positive eugenics and "negative measures for race improvement."[59] In 1912, New York University offered a course titled The Family and Eugenics, while Columbia University and Barnard College each offered a eugenics-based course titled Biology and Vital Relations of the Human Growth. Other New York colleges that taught eugenics were Adelphi (Biology), Cornell (Theory of Evolution), Colgate (Development and Heredity), Farmingdale (Dairying and Animal Husbandry), Fordham (Philosophical Biology), Syracuse (Zoology, Mental Health and Hygiene) and Vassar (Zoology and Botany).[60]

By 1914, forty-four colleges either offered complete courses in eugenics or had eugenics-based lectures as a significant part of their courses.[61] Sixteen of these colleges taught eugenics in their zoology departments, while eleven of them taught the subject in their biology and sociology departments. Cornell University offered a course called Genetics and Eugenics. Several eugenics-based courses were also offered at the College of William and Mary in Williamsburg, Virginia, where from 1913 to 1919, a course titled Evolution and Heredity provided "a series of lectures dealing with the broader aspects of biology and the social applications of biological principles."[62] Other courses in the fields of genetics or evolution provided eugenics-based material to students as well.

In the West, eugenics was a regularly offered course in the biology department at San Francisco State University from 1926 to 1951.[63] According to the course description, the course presented "the study of the facts and the problems of human heredity and possibility of race betterment." It is believed that other similar courses were offered in western states on the subject of human genetics. Overall, there was no escaping the presence of eugenics in hundreds of colleges throughout the United States.

In tandem with the many eugenics courses taught across the nation, many academic textbooks and treatises were written on the topic. In 1918, Paul Popenoe, a leading eugenicist from California, and Roswell Hill Johnson from the University of Pittsburgh coauthored a book titled *Applied Eugenics*. After racing through four printings in six years, the book became a leading textbook in the eugenics world.[64] Other popular eugenics books included *The Right to Be Well Born* (1917) by William Earl Dodge Stokes and *Eugenics and Mental Life* (1911) by Robert Yerkes, whose influence led to the administration of intelligence tests to nearly two million U.S. military officers.[65] In 1920, William McDougall published *The Group Mind* and followed up one year later with *Is America Safe For Democracy?* Both books embraced eugenics, espoused numerous racist beliefs and advocated for the sterilization of those deemed to be defective.

Influential professors at Harvard also wrote many books on eugenics. In 1919, Harvard botanist Edward M. East published *Inbreeding and Outbreeding: Their Genetic and Sociological Significance*. In this book, he railed against Blacks, Jews, Italians and Asians and warned that race mixing would be harmful for the White race. In 1923, he released *Mankind at the Crossroads*, in which he wrote, "Eugenics is sorely needed; social progress without it is unthinkable."[66] In 1927, William E. Castle, a professor of zoology at Harvard University, published a leading textbook on eugenics titled *Genetics and Eugenics: A Text-Book for Students of Biology and a Reference Book for Animal and Plant Breeders*. In the book, Castle wrote, "From the viewpoint of the superior race there is nothing to be gained by crossing with an inferior race." The book was published by Harvard University Press and bore Harvard's "veritas" seal on the title page.

In 1920, prominent Harvard eugenicist Lothrop Stoddard released the bestselling book *The Rising Tide of Color Against White World Supremacy*. The book went through fourteen printings in its first three years, drew lavish praise from President Warren G. Harding and was vaguely referenced in the novel *The Great Gatsby*. Finally, Charles Davenport himself was squarely among this Harvard academic scene at the time. In 1911, he published *Heredity in Relation to Eugenics*, which became the standard book for eugenics courses at colleges and universities nationwide and was cited by more than one-third of high school biology textbooks of the era.[67]

Clearly, the infusion of eugenics into American academic institutions was a potent stratagem aimed at further legitimizing the pseudoscience and solidifying it into the fabric of the country for both present and future generations.

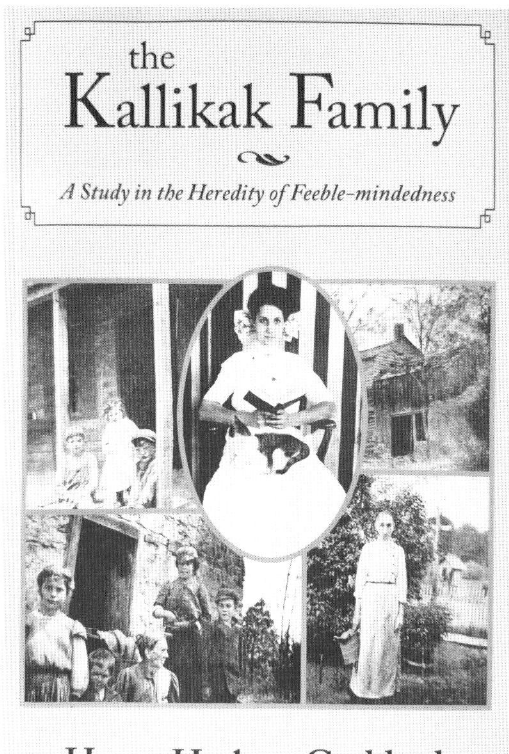

The Kallikak Family: A Study in the Heredity of Feeble-mindedness by Henry Herbert Goddard. Released in 1912, the book was a bestseller among eugenic propaganda. *Photo courtesy of Mark A. Torres.*

THE EUGENICS MOVEMENT ALSO benefited from a large number of commercially published books written on the topic. As opposed to the literature written for students in academic settings, these books targeted the public at large. They were written by authors who were eugenicists themselves with a concerted interest in promoting literature that reinforced their belief that heredity was based on physical characteristics along with social behavior and temperament.

One of these books was *The Kallikak Family: A Study in the Heredity of Feeble-mindedness*. Authored by Henry Goddard in 1912, *The Kallikak Family* was a best-selling book that further solidified Goddard's view that feeblemindedness was an inheritable trait that contributes to crime, poverty and other social problems. It presented a cautionary tale about the dangers of procreation among the so-called defectives within society. Goddard's work at the Vineland Training School for Feeble-Minded Boys and Girls had already earned him a powerful reputation in the field of eugenics. His book drew high praise from eugenicists worldwide.

Photograph taken in 1915 of Ulster County, New York, as part of the fieldwork conducted by Arthur Estabrook for his book *The Jukes of 1915. Photo courtesy of Arthur Estabrook Papers, M.E. Grenander Special Collections & Archives, University at Albany, SUNY.*

In 1877, Richard Dugdale, an executive in the New York prison system, conducted a study of the ancestry of a large group of criminals, prostitutes and social misfits through seven generations of a single set of parents in Ulster County, New York.[68] In his book, titled *The Jukes: A Study in Crime, Pauperism, Disease, and Heredity*, Dugdale attributed the troubles of this family to the environment they lived in. This element was largely ignored, and eugenicists solely seized on the overall problems identified in the book to call for corrective eugenic measures. Indeed, this tunnel vision approach was a foundation of American eugenics.

In 1916, Arthur Estabrook, a field officer with the Eugenics Record Office, followed up on Dugdale's work with his own book, titled *The Jukes of 1915*.[69] In it, Estabrook ignored the environmental factors explored by Dugdale and blamed the struggles of the latter-day family solely on poor hereditary traits. Estabrook also released a comprehensive report of numerous subjects he studied and compiled into a fictitious family, who left Massachusetts for a region in upstate New York. Charles Davenport lauded the book, summarizing the Jukeses as a family beset with "feeblemindedness, indolence, licentiousness, and dishonesty."[70]

Other books written by Estabrook included *The Dack Family* (of central Pennsylvania) and *The Hill Folk Report*, which detailed a study of two family trees from a small Massachusetts town in "the fertile valley of the New

England hills."[71] In a customary tactic repeatedly utilized by eugenicists to strengthen public support for eugenics, these works detailed the heavy financial burden placed on the state to provide care for so-called state wards, which from 1888 to 1911 was estimated to be nearly $46,000.

Estabrook also coauthored a book with Charles Davenport titled *The Nam Family: A Study in Cacogenics*. This book focused on a rural community in New York founded by a White Dutch patriarch and an "Indian princess." According to the authors, this union led to a lineage with unusual levels of alcoholism, lack of ambition and general dysfunction. In summation, they wrote, "No state can afford to neglect such a breeding center of feeblemindedness, alcoholism, sex-immorality, and infanticide as we have here. A rotten apple can infect the whole barrel of fruit." Both Estabrook and Davenport openly called for the sterilization of the entire Nam community.

FEW FORMS OF MEDIA have ever been as influential as the moving picture. Thus, it is no surprise that cinematic productions were another useful vehicle for spreading eugenic ideals. In 1917, the highly anticipated film *The Black Stork* was released. The story centers on a fictional couple who are counseled by a doctor to not have a child because it would likely be defective. Dr. Harry Haiselden, the Illinois doctor who was hailed as a hero by eugenicists across the nation for allowing a deformed child to die without medical care, was cast in a starring role. In the film, a newborn child with a birth defect is allowed to die, and when it does, the dead child levitates into the arms of Jesus Christ. The eugenics propaganda film was hugely successful and played regularly in theaters throughout the country for more than a decade.

In 1934, *Tomorrow's Children*, another eugenics-based film, presented a cautionary tale about good intentions gone awry. The film centers on a seventeen-year-old woman named Alice Mason who is slated for sterilization because her family comprises drunkards, cripples and the feebleminded. She is spared the procedure only because of the revelation that she was adopted.

In 1937, despite the ongoing atrocities being committed in Germany under the banner of eugenics, Harry Laughlin of the Eugenics Record Office and the Eugenics Research Association became a distributor of two copies of a Nazi eugenics propaganda film titled *Erbkrank* or *The Hereditarily Defective*. The film was made by German eugenicists to address the problems of "hereditary degeneracy in the fields of feeble-mindedness, insanity, crime, hereditary disease and inborn deformity." Laughlin was particularly active in promoting the film, loaning it to high schools in New York and New

Jersey, as well as to welfare workers in Connecticut and to the Society for the Prevention of Blindness.[72] In a letter to Wickliffe Preston Draper, Laughlin gleefully stated, "You will be interested to know that the moving picture film 'Eugenics in Germany' has proven very popular with senior high school students. Up to date, the film has been loaned 28 times."[73]

ALL MOVEMENTS REQUIRE THE support and participation of people with strong public influence. While many influential Americans were eager to promote eugenics, there were few greater endorsements than that of the president of the United States of America. In fact, "every president from Theodore Roosevelt to Herbert Hoover was a member of a eugenics organization, publicly endorsed eugenic laws, or signed eugenic legislation without voicing opposition."[74]

On January 3, 1913, Theodore Roosevelt wrote a heartfelt letter to Charles Davenport to express his fervent support of eugenics. He wrote that "society has no business to permit degenerates to reproduce their kind....It is really extraordinary that our people refuse to apply to human beings such elementary knowledge as every successful farmer is obliged to apply to his own stock breeding." In closing, Roosevelt added, "Someday we will realize that the prime duty of the good citizen of the right type to leave his blood behind him in the world; and that we have no business to perpetuate citizens of the wrong type."[75] Roosevelt spent a great deal of time in his "summer White House" in Sagamore Hill, Oyster Bay, which was a mere six miles from the ERO facility in Cold Spring Harbor. Nearly two decades later, in August 1933, Anna Eleanor Roosevelt (wife of President Franklin D. Roosevelt) met with Vassar students at the Institute of Eugenics in Poughkeepsie, New York, lending further support for the movement.[76]

Eugenics was also preached in religious institutions. In 1927, the American Eugenics Society created a sermon contest in which preachers across the nation crafted sermons praising the values of eugenics. One such sermon read: "Progress is the law of all life. Not to progress is unnatural and abnormal. Among the principal factors contributing to world progress is Modern Science, and second to none among the Sciences is the comparatively new Science called Eugenics."[77]

Many other prominent Americans supported eugenics, including John Harvey Kellogg (1852–1943), an American businessman, inventor and doctor with a strong interest in progressive causes, among them eugenics. Kellogg, along with his brother, founded the Kellogg company, which developed corn flakes and other breakfast products. John Kellogg was a board member of a health sanitarium in Battle Creek, Michigan, and a staunch proponent of digestive health and well-being. He was also a staunch ally of Charles Davenport and a full-fledged eugenicist.

In 1906, Kellogg established a eugenics and racial hygiene organization called the Race Betterment Foundation. In 1914, he organized the First Race Betterment Foundation Conference in Battle Creek, Michigan, with the stated purpose of establishing the foundations for the creation of a super race. In promoting the event, Kellogg stated, "We have wonderful new races of horses, cows, and pigs. Why should we not have a new and improved race of men?" He believed that this improved race should be "white races of Europe."[78]

Charles Davenport attended the lavish conference. In a speech, he touted his ability to work with state institutions to gather massive amounts of eugenics-based information. He told the audience, "We have found that a large proportion of the feeble-minded, the great majority of them, are such because they belong to defective stock." Harry Laughlin was also in attendance. With fiery passion, he declared, "To purify the breeding stock of the race at all costs is the slogan of eugenics." He also promoted a three-pronged eugenic approach that included sterilization, mass incarceration and sweeping immigration-restrictive measures. The event was wildly successful and helped promote eugenics to a national audience.

A banquet for physicians and delegates to the National Conference on Race Betterment hosted by Harvey Kellogg at the Battle Creek Sanitarium in Battle Creek, Michigan, on January 10, 1914. *Photo courtesy of the American Philosophical Society.*

Some prominent Americans from unlikely backgrounds also openly expressed support for eugenics. In 1903, the sociologist and civil rights activist W.E.B. Du Bois wrote an essay titled "The Talented Tenth" in which he argued that one out of every ten Black men had the ability to become educated and thus had an obligation to become leaders of their communities. "The Negro race," Du Bois wrote, "like all other races, is going to be saved by exceptional men. The problem of education, then, among Negroes must first of all deal with the Talented Tenth; it is the problem of developing the Best of this race that they may guide the Mass away from the contamination and death of the Worst."[79] Du Bois was also a strong advocate for birth control.

Helen Keller lost her sight and hearing at a very young age. She went on to become an author, disability rights advocate and political activist. She was also a supporter of eugenics and once wrote, "Our puny sentimentalism has caused us to forget that a human life is sacred only when it may be of use to itself and the world."[80] Du Bois was a man of color and Keller a woman with disabilities. They each represented segments of the population that were specifically targeted by eugenicists and subjected to harsh treatment. Whether they truly believed in the pseudoscience or their actions were the product of self-preservation may never be known, but their support for eugenics certainly indicates its widespread influence at the time.

In a continuous effort to promote the movement, eugenicists participated in several elaborate national and international events. In August 1915, San Francisco hosted a fair called the Panama-Pacific International Exposition. This fair promoted numerous educational, scientific and technological exhibits on over six hundred acres along two and a half miles of waterfront property. Between February and December 1915, over eighteen million people visited the fair.[81] Special events were held daily, and souvenirs were available for visitors.

Among the many attractions at this event was a carefully constructed exhibit called Race Betterment: A Popular Non-Sectarian Movement to Advance Life Saving Knowledge. The entire eugenics exhibit was planned through the collaboration of Harvey Kellogg and Charles Davenport, had six booths and was housed in the Palace of Education building. On entry, visitors were greeted by large plaster casts of Atlas, Apollo and Venus. On display were numerous charts depicting the degenerative effects of alcohol, rules for healthy living and dieting, a list of eugenics organizations across the nation and an assortment of medical instruments used by doctors to evaluate the physiological and biological capacities of humans through various age groups.[82] One item that drew particular attention from visitors was the so-called New Laughlin Gyotometer. This eugenic device, believed to have been invented by Harry Laughlin at the ERO, was used to determine "various hereditary results from parent combinations."

The eugenics exhibit at the Panama-Pacific International Exposition was wildly successful and very popular. According to event organizers, more than one thousand visitors inspected the collection each day, many of whom returned twice a day and others visiting six or seven times per day. Participation at this event undoubtedly provided a national audience for the pseudoscience of eugenics.

The eugenics-based exhibit titled "Race Betterment: A Popular Non-Sectarian Movement to Advance Life Saving Knowledge" at the Panama-Pacific International Exposition held in San Francisco, California, in 1915. Organizers estimated that more than eighteen million visitors attended this event between February and December 1915, with many visiting the eugenics exhibit multiple times. *Photo courtesy of the Edward A. Rogers collection of Cardinell-Vincent Company and Panama-Pacific International Exposition photographs, BANC PIC 2015.013:11182.5—NEG, The Bancroft Library, University of California, Berkeley.*

Eugenicists also hosted several international conferences known as the International Eugenics Congresses. These events served two primary purposes. The first was to present periodic updates on the status of eugenics. The second was to keep wealthy donors and members of the public enthusiastically engaged in the movement. Scientists from all over the world were invited to share their work and present it to members of the public. In total, three international conferences were held between 1912 and 1932, and they were all presided over by Charles Davenport; Leonard Darwin, son of Charles Darwin; and Henry Fairfield Osborn, president of the Museum of Natural History in New York.

The first International Eugenics Congress was held in London, England, from July 24 to 30, 1912. Approximately four hundred delegates gathered at the Hotel Cecil in London. Scientists presented various research papers, shared ideas on how to continue promoting eugenics and established organizations to pursue eugenic ideals.[83] Sybil Neville-Rolfe, a women's rights advocate in the United Kingdom, served as the honorary secretary. Committees from the United States, the United Kingdom, Belgium, France, Germany and Italy were in attendance.

Over one hundred scientific presentations were on display, including portraits, charts and research papers. Two of the exhibits were titled "The Cause of the Inferiority of Physical and Mental Characters in the Lower Social Classes" and "Heredity and Eugenics in Relation to Insanity." Several organizations sent representatives to attend the conference, including the University of Cambridge, the University of Oxford, the American Association for the Advancement of Sciences and the American Philosophical Society.

Thirty-seven members were bestowed with the honorary title of vice president, including Alexander Graham Bell and Winston Churchill. A Committee on Sterilization was also established, which created a written review of the committee's first report, that was drafted and published by Harry Laughlin in 1914. The event was hailed as a victory for the eugenics movement.

The second International Eugenics Congress took place from September 22 to 28, 1921. This event was held at the American Museum of Natural History in New York City, with Henry Fairfield Osborn presiding. Alexander Graham Bell served as honorary president, and Sybil Neville-Rolfe returned to serve as honorary secretary. Charles Davenport, Henry Osborn and Leonard Darwin delivered opening remarks.

Representatives from North and South America and Europe presented over one hundred exhibits. Even though there were no representatives from

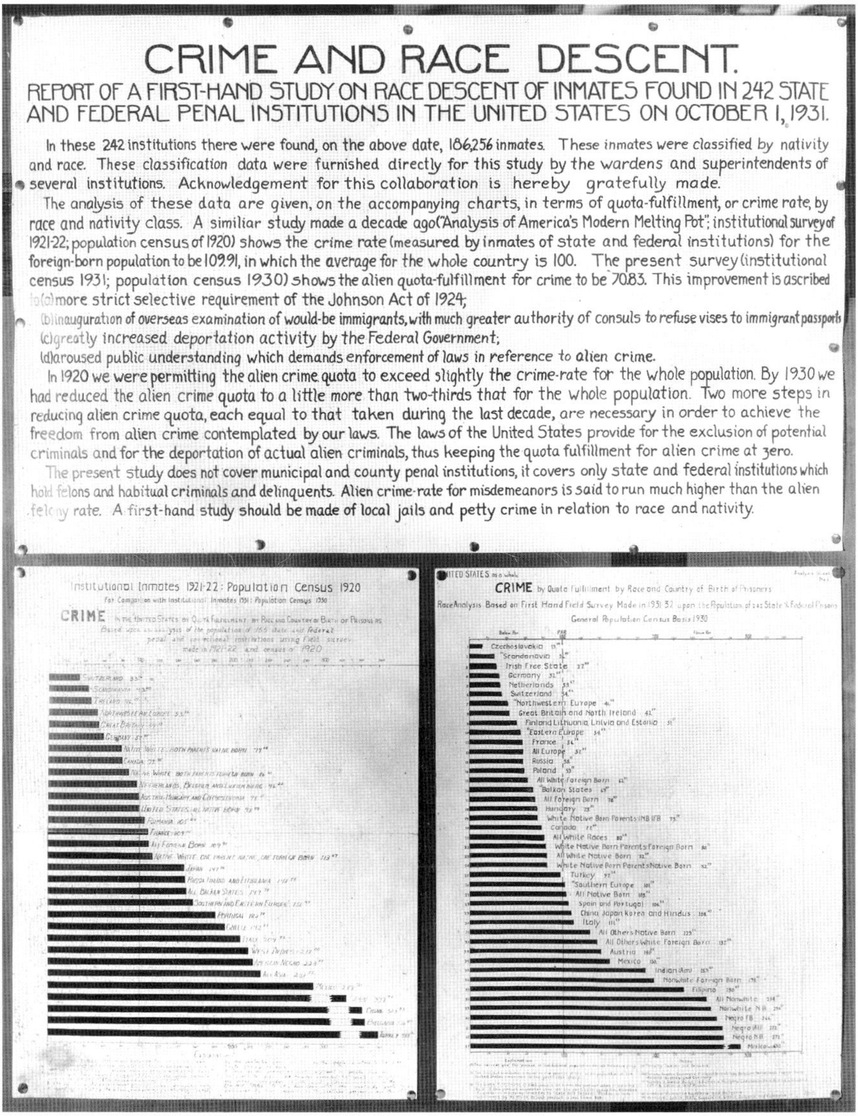
CRIME AND RACE DESCENT.
REPORT OF A FIRST-HAND STUDY ON RACE DESCENT OF INMATES FOUND IN 242 STATE AND FEDERAL PENAL INSTITUTIONS IN THE UNITED STATES ON OCTOBER 1, 1931.
In these 242 institutions there were found, on the above date, 186,256 inmates. These inmates were classified by nativity and race. These classification data were furnished directly for this study by the wardens and superintendents of several institutions. Acknowledgement for this collaboration is hereby gratefully made.
The analysis of these data are given, on the accompanying charts, in terms of quota-fulfillment, or crime rate, by race and nativity class. A similiar study made a decade ago("Analysis of America's Modern Melting Pot"; institutional survey of 1921-22; population census of 1920) shows the crime rate (measured by inmates of state and federal institutions) for the foreign-born population to be 109.91, in which the average for the whole country is 100. The present survey (institutional census 1931; population census 1930) shows the alien quota-fulfillment for crime to be 70.83. This improvement is ascribed to (a) more strict selective requirement of the Johnson Act of 1924;
(b) inauguration of overseas examination of would-be immigrants, with much greater authority of consuls to refuse vises to immigrant passports
(c) greatly increased deportation activity by the Federal Government;
(d) aroused public understanding which demands enforcement of laws in reference to alien crime.
In 1920 we were permitting the alien crime quota to exceed slightly the crime-rate for the whole population. By 1930 we had reduced the alien crime quota to a little more than two-thirds that for the whole population. Two more steps in reducing alien crime quota, each equal to that taken during the last decade, are necessary in order to achieve the freedom from alien crime contemplated by our laws. The laws of the United States provide for the exclusion of potential criminals and for the deportation of actual alien criminals, thus keeping the quota fulfillment for alien crime at zero.
The present study does not cover municipal and county penal institutions, it covers only state and federal institutions which hold felons and habitual criminals and delinquents. Alien crime-rate for misdemeanors is said to run much higher than the alien felony rate. A first-hand study should be made of local jails and petty crime in relation to race and nativity.
Institutional Inmates 1921-22: Population Census 1920
CRIME
CRIME by Quota Fulfillment by Race and Country of Birth of Prisoners
General Population Census Basis 1930

A report titled *Crime and Race Descent*, which purported to link crimes committed by U.S. inmates at various institutions by race and place of birth in comparison to other countries, at the Third International Eugenics Congress in 1932. *Photo courtesy of the Truman State University, Pickler Memorial Library, Special Collections and Museums.*

A display about Charles Darwin at the Third International Eugenics Congress in 1932.
Photo courtesy of the Truman State University, Pickler Memorial Library, Special Collections and Museums.

This page and opposite bottom: Eugenics exhibits at the Third International Eugenics Congress in 1932. *Photos courtesy of the Truman State University, Pickler Memorial Library, Special Collections and Museums.*

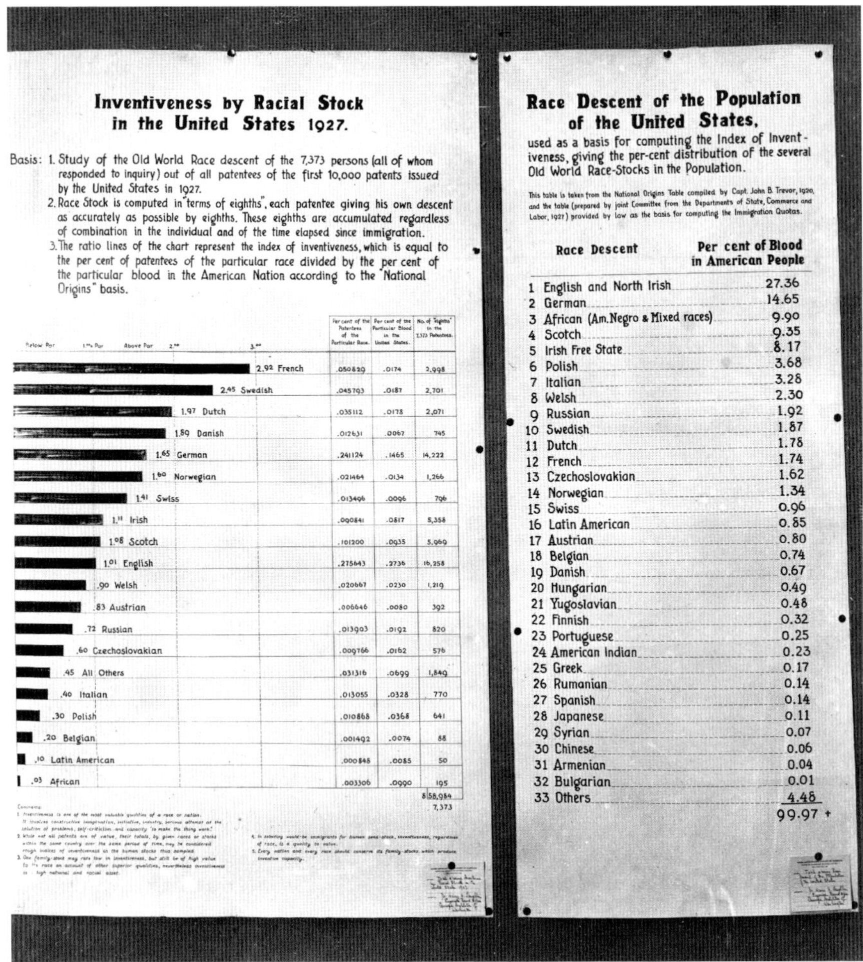

Germany due to international tensions after the First World War, German eugenicists continued to collaborate with their U.S. counterparts at the ERO. By this time, there was no doubt about the United States' dominance in the field of eugenics, as forty-one of the fifty-three scientific papers were produced by U.S. eugenicists.[84] Topics discussed included a study of the so-called mountain people of Kentucky and the work on heredity blindness by Lucien Howe. Some of the other presentations included "The Problem of Negro-White Intermixture and Intermarriage" and "Preventative Eugenics: The Protection of Parenthood from the Racial Poisons." Event organizers claimed that the public exhibit drew between five thousand and ten thousand visitors. Field trips were organized for international attendees to visit the ERO in Cold Spring Harbor.

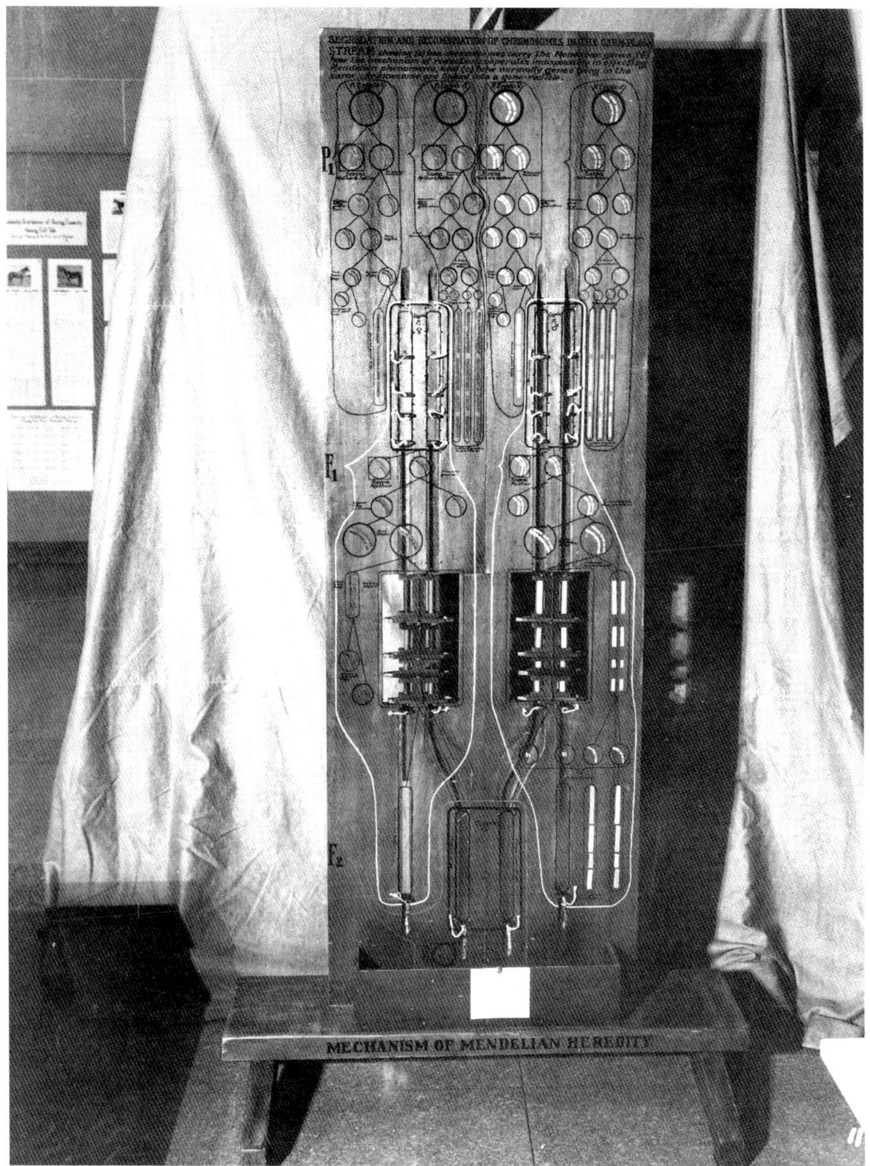

Opposite: A eugenics exhibit titled "Inventiveness by Racial Stock in the United States 1927" at the Third International Eugenics Congress in 1932. *Photo courtesy of the Truman State University, Pickler Memorial Library, Special Collections and Museums.*

Above: A eugenic device called the "Mechanism of Mendelian Heredity" on display at the Third International Eugenics Congress in 1932. The device was purportedly designed to display the "Segregation and Recombination of Chromosomes in the Germ-Plasm Stream." *Photo courtesy of the Truman State University, Pickler Memorial Library, Special Collections and Museums.*

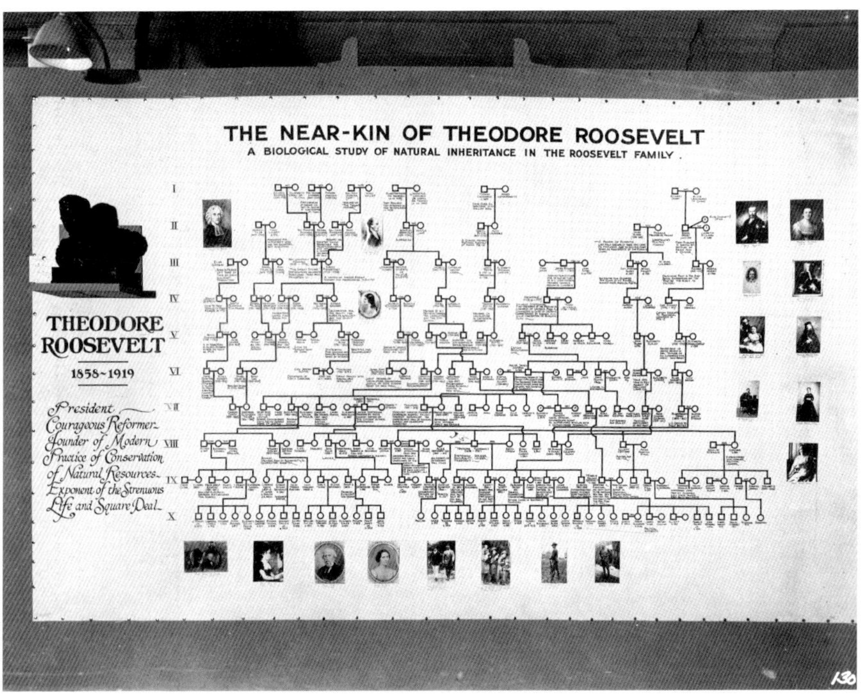
THE NEAR-KIN OF THEODORE ROOSEVELT
A BIOLOGICAL STUDY OF NATURAL INHERITANCE IN THE ROOSEVELT FAMILY.
THEODORE ROOSEVELT
1858~1919
President
Courageous Reformer
Founder of Modern Practice of Conservation of Natural Resources
Exponent of the Strenuous Life and Square Deal
I
II
III
IV
V
VI
VII
VIII
IX
X

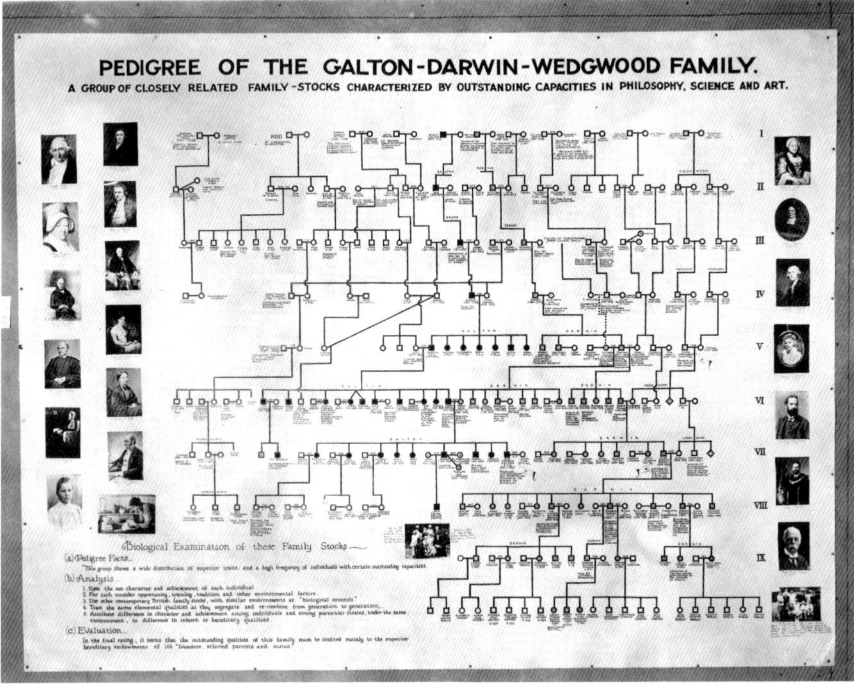
PEDIGREE OF THE GALTON-DARWIN-WEDGWOOD FAMILY.
A GROUP OF CLOSELY RELATED FAMILY-STOCKS CHARACTERIZED BY OUTSTANDING CAPACITIES IN PHILOSOPHY, SCIENCE AND ART.
I
II
III
IV
V
VI
VII
VIII
IX
GALTON
DARWIN
WEDGWOOD
Biological Examination of these Family Stocks
(a) Pedigree Facts
This group shows a wide distribution of superior traits, and a high frequency of individuals with certain outstanding capacities.
(b) Analysis
1. Rate the net character and achievement of each individual
2. For each consider opportunity, training, tradition and other environmental factors.
3. Use other contemporary British family stocks, with similar environments as "biological controls"
4. Trace the same elemental qualities as they segregate and re-combine from generation to generation.
5. Attribute difference in character and achievement among individuals and among particular strains, under the same environment, to difference in inborn or hereditary qualities.
(c) Evaluation
In the final rating, it seems that the outstanding qualities of this family must be credited mainly to the superior hereditary endowments of its "founders, selected parents and mates"

Eugenics Congress Announcement

Number 2. The Exhibit

Education Hall, American Museum of Natural History
New York City, August 22–September 22, 1932

Sir Francis Galton, Founder of the Science of Eugenics
This picture, taken in 1895, showing Galton aged 73, is, with permission, reproduced from Pearson's "Life of Galton," vol. II

"The term National Eugenics is here defined as the study of the agencies, under social control, that may improve or impair the racial qualities of future generations, either physically or mentally." *Paragraph from Galton's will, 1906*

Third International Congress of Eugenics

Honorary Presidents, Leonard Darwin and Henry Fairfield Osborn
President, Charles B. Davenport

New York City, August 21–23, 1932

Opposite, top: A eugenics exhibit of the pedigree chart titled "The Near-Kin of Theodore Roosevelt" on display at the Third International Eugenics Congress in 1932. *Photo courtesy of the Truman State University, Pickler Memorial Library, Special Collections and Museums.*

Opposite, bottom: A eugenics exhibit of the pedigree chart titled "The Galton-Darwin-Wedgewood Family" on display at the Third International Eugenics Congress in 1932. *Photo courtesy of the Truman State University, Pickler Memorial Library, Special Collections and Museums.*

Above: A eugenics exhibit titled "Study of the South African Negro" on display at the Third International Eugenics Congress in 1932. *Photo courtesy of the Truman State University, Pickler Memorial Library, Special Collections and Museums.*

Left: A promotional pamphlet for the Third International Eugenics Congress in New York City, August 21 to 23, 1932, with a photograph of Sir Francis Galton on the cover. *Photo courtesy of the American Philosophical Society.*

The event benefited from funding from the Carnegie Institute, which donated $2,000 for travel expenses to and from Europe, and from Mary Harriman, who donated $2,500. Prior to the close of the conference, two new eugenics organizations were established. The first was the American Eugenics Society, which was to be headed by economist Irving Fisher. The International Eugenics Commission was also established, with Leonard Darwin to serve as chairman and Henry Osborn to serve as vice chairman. This committee was later renamed the International Federation of Eugenics Organization (or IFEO).[85]

The third and final International Eugenics Congress was also held at the American Museum of Natural History in New York City, from August 21 to 23, 1932. The event was dedicated to Mary Harriman for her staunch support of the eugenics movement. Charles Davenport served as president, and Harry Laughlin served as secretary and exhibit chairman. Henry Osborn provided the keynote speech, titled "Distinction Between Birth Control and Birth Selection," a call for eugenics to cure the social ills of overpopulation and underemployment and for reproduction of the fittest persons in society.

A variety of topics were discussed, and achievements in eugenics were announced. More than sixty scientific papers on a variety of eugenics-based topics and other updates were on display. Some of the public exhibits included pedigree charts of prominent families like those of Galton and Darwin, George Washington, Abraham Lincoln and Theodore Roosevelt. Charts with anthropological measurements were also displayed, including photos of skeletons of Native Americans and people from Africa and Asia. These were accompanied by statistics on crime, disease and birth and death rates. Once again, a field trip to the ERO was organized.

On the final day of the conference, the IFEO elected Ernst Rudin president of the organization. Rudin was a eugenicist who helped lead a program in Germany that would ultimately lead to the sterilization of four hundred thousand people against their will.

Unlike the first two events, this conference was poorly attended, and the event received poor reviews from the press. Davenport blamed the negative press on political tensions with Germany and the global economic depression. Nevertheless, the three International Eugenics Congresses allowed a gathering of scientific minds to coalesce around their common and global interests in eugenics. They also entertained donors and allowed access to the public with the intended purpose of expanding interest in eugenics. The last two conferences, held in New York City, would have been considered star-studded events in the world of eugenics.

TO INCREASE THE POPULARITY of the movement, eugenicists organized events at various state fairs, which included public contests in which local families could compete to show their superior traits. Known as Fitter Families contests, these competitions were orchestrated by the American Eugenics Society (AES). The first such event was held in Topeka, Kansas, in 1920.[86] Other states like Arkansas, Michigan and Texas held similar fairs. They were so popular that nearly ten were held each year, and by the end of the decade, requests from more than forty sponsors seeking guidance to hold the fairs were made annually. By 1930, the AES had drafted proposed itineraries for Fitter Families contests at fairs in Connecticut, Massachusetts, Rhode Island and Vermont.[87]

The Fitter Families competitions were typically held in "human stock" sections of state fairs. A brochure for one of these events proclaimed:

> *The time has come when the science of human husbandry must be developed, based on the principles now followed by scientific agriculture, if the better elements of our civilization are to dominate or even survive.*[88]

Fitter Families contest, Kansas Free Fair. *Photo courtesy of the American Philosophical Society.*

"Eugenic and Health Exhibit" at the Kansas Free Fair, 1929. *Photo courtesy of the American Philosophical Society.*

Participation in these events was open to any healthy family that could provide pedigree information. During the contest, each family member had to undergo a eugenics-based medical examination, a psychological evaluation and an intelligence test. The winners were typically categorized in either small, average or large family sizes. At the 1924 Kansas Free Fair, winning families were awarded a Governor's Fitter Family Trophy, which was presented which much pageantry.

These Fitter Family contests were adorned with a wide array of exhibits aimed to further promote eugenics. These included the usage of stuffed black and white guinea pigs arranged in such a manner as to depict the inheritance of color coats from generation to generation.[89] Elaborate charts were also displayed illustrating the Mendelian laws of inheritance for human beings. For instance, some of these charts depicted a cross between two "pure" parents who would produce "normal" children. Conversely, other charts showed a cross between two "abnormal" parents who would produce "abnormal" children. One exhibit prominently stated, "How long

A eugenics propaganda poster that reads: "Eugenics: Like a tree, eugenics draws its materials from many sources and organizes them into an harmonious entity." *Photo courtesy of the American Philosophical Society.*

are we Americans to be so careful for the pedigree of our pigs and chickens and cattle—and then leave the ancestry of our children to chance or to 'blind' sentiment?"[90]

Fitter Family contests were widely popular. They provided opportunities for families to participate in the fiercely competitive eugenics-based contests. As they did so, eugenics became further entrenched in the heartland of America as a legitimate and outright necessary science for mankind.

THE AMERICAN EUGENICS MOVEMENT did not only thrive as a singular cause. Given its enormous popularity, it also tended to blend with a vast number of social reform and health care movements that had also arisen at the time. Child welfare, prison reform, improved education and immigrant rights were all causes led by charitable organizations that eugenicists sought to infiltrate. Among the most polarizing of these causes was perhaps the birth control movement and the effort to assist women in making independent choices about their own pregnancies, largely driven by the ever controversial Margaret Sanger.

In the early twentieth century, many women in the United States were relegated to being dutiful wives and pressured to bear children regardless of their health or quality of life.[91] Women were also treated as second-class citizens and denied the right to vote. Margaret Sanger began to fight for women who had little choice in when or how often they became pregnant in a society so strict that even discussing birth control was likened to distributing pornographic material.[92]

Margaret Higgins Sanger was born in Corning, New York, on September 14, 1879. She was the sixth of eleven children. As a young adult, she worked feverishly as a nurse in the slums of Manhattan and Brooklyn and encountered numerous unwanted pregnancies. In her autobiography, she recounted one night in 1912 when she was called to assist Jake and Sadie Sachs.[93] The young couple already had three children and were completely ignorant about exercising reproductive controls. Just months earlier, Sadie performed a dangerous self-induced abortion. Now, the young woman was pregnant again and in a life-threatening condition. Sanger raced to the apartment and found Sadie in a comatose state. Just ten minutes later, the young woman died. This tragic event catapulted Sanger into a life of advocacy for birth control, a cause she considered to be an integral part of population control.

Through her life of advocacy, Sanger developed many enemies who, even after her death, sought to undermine and discredit her work. She has often been labeled racist or anti-Semitic. Undeniably, Margaret Sanger was a staunch supporter of eugenics, and she often used her platform to legitimize the idea for mass sterilization and incarceration of so-called defectives, whom she referred to as "unfit" and "weeds" that should be "exterminated." She also endorsed the need for strict immigration restrictions.[94] In her 1922 book *The Pivot of Civilization*, Sanger wrote:

> *The emergency problem of segregation and sterilization must be faced immediately. Every feeble-minded girl or woman of the hereditary type, especially of the moron class, should be segregated during the reproductive period. Otherwise, she is almost certain to bear imbecile children, who in turn are just as certain to breed other defectives. The male defectives are no less dangerous. Segregation carried out for one of two generations would give us only partial control of the problem. Moreover, when we realize that each feeble-minded person is a potential source of an endless progeny of*

> *defect, we prefer the policy of immediate sterilization, of making sure that parenthood is absolutely prohibited to the feeble-minded.*[95]

One of the eugenicists' consistent strategies was to use economic data to appeal to the wealthy and the middle class to cease their charitable efforts to help the so-called defectives in society. Sanger echoed this sentiment and regularly railed against charitable efforts to uplift the poor, writing,

> *Organized charity itself is the symptom of a malignant social disease. Those vast, complex, interrelated organizations aiming to control and to diminish the spread of misery and destitution and all the menacing evils that spring out of this sinisterly fertile soil, are the surest sign that our civilization has bred, is breeding, and is perpetuating constantly increasing numbers of defectives, delinquents, and dependents. My criticism, therefore, is not directed at the failure of philanthropy, but rather at its success.*[96]

Margaret Sanger always aligned her birth control movement with eugenics. When she founded the American Birth Control League, she set forth a list of goals for the organization. The fourth item on that list was "sterilization of the insane and feebleminded and the encouragement of this operation upon those affected with inherited or transmissible diseases."[97] She also believed that 70 percent of the American population had the intellect of a fifteen-year-old or younger. In her book *Woman and the New Race*, she stated,

> *Many, perhaps, will think it idle to go farther in demonstrating the immorality of large families, but since there is still an abundance of proof at hand, it may be offered for the sake of those who find difficulty in adjusting old-fashioned ideas to the facts. The most merciful thing that the large family does to one of its infant members is to kill it.*[98]

Sanger surrounded herself with White supremacists who were inextricably linked to the eugenic movement. These included Lothrop Stoddard, who wrote the 1920 book *The Rising Tide of Color Against White World Supremacy*. Stoddard compared the downtrodden and so-called defectives to bacteria and wrote, "We now know that men are not, and never will be, equal."[99] Sanger's relationship with Stoddard was by no means tangential. In 1920, she invited him to join the board of directors of her American Birth Control League, a position he accepted and held for years. Sanger was also a colleague of Irving Fisher, an economics professor at Yale, early board member of the ERO and staunch leader of the Eugenics Research Association and the American Breeders Association. Speaking at the Second National Congress on Race Betterment held in San Francisco in 1915, Fisher openly declared, "You have not any idea unless you have studied this subject mathematically, how rapidly we could exterminate this contamination if we really get at it, or how rapidly the contamination goes on if we do not get at it." Sanger eventually appointed Fisher to serve as vice president of the National Committee for Federal Legislation on Birth Control, a lobbying organization for her group. Another colleague of Margaret Sanger was Henry Pratt Fairchild, who served as a lead organizer. In 1926, Fairchild's book *The Melting Pot Mistake* was released, in which he freely expressed his anti-immigrant rants and compared ethnic minorities to bacteria.[100]

Despite her close affiliation with some of the leading eugenicists of the time, Margaret Sanger was never truly embraced by or officially welcomed into the eugenics movement. Her attempts to officially link her organization with Charles Davenport and the Eugenics Research Association were continuously rebuffed. Even though she fully embraced negative eugenics, which included the elimination of the unfit through mass sterilization and segregation, Sanger refused to accept *constructive eugenics*, a term used to describe the eugenic ideal that superior families should not be hindered from having many children.[101] On this point, she wrote,

> *But in its so-called "constructive" aspect, in seeking to reestablish the dominance of the healthy strain over the unhealthy, by urging an increased birth-rate among the fit, the eugenicists really offer nothing more farsighted than a "cradle competition" between the fit and the unfit. They suggest in very truth, that all intelligent and respectable parents should take as their example in this grave matter of child-bearing the most irresponsible elements in the community.*[102]

In the true Malthusian spirit, whereby overall human population control was critical, Margaret Sanger believed that all families, regardless of whether they were deemed superior or unfit, should practice intelligent birth control, and it was this belief that ultimately thwarted her full acceptance into the American eugenics movement. Nevertheless, by forging relationships with many leading eugenicists, Sanger had aligned the American birth control movement with eugenics, as did many other social movements of the time.

Chapter 5

"SCIENTIFIC RACISM" AND THE ANTI-IMMIGRATION MOVEMENT

Can we build a wall high enough around this country to keep out these cheaper races?
—Charles Davenport to Madison Grant, 1920

The ability to make eugenics a household name throughout the country was a calculated effort fueled by a great deal of national strife in the United States during the latter part of the nineteenth century. Between 1880 and 1920, more than twenty million immigrants flooded into the United States, mainly from Europe. More than eight million of those arrived between 1900 and 1909.[103] After the First World War, millions of Americans lost their jobs in industries that either flourished only during wartime production or during the many ongoing labor strikes. According to the 1920 census, for the first time in the country's history, the majority of the population had shifted from rural to urban areas. Race riots and violence intensified. Social fears, ethnic strife and economic hardship stimulated plans for reforms. Eugenicists did not hesitate to exploit this turbulent climate.

The eugenics movement was already receiving anti-immigration support from prominent Americans like Alexander Graham Bell, the prominent inventor and engineer who would later serve as president of the second International Eugenics Congress in New York. In 1920, Bell wrote to Davenport suggesting that more should be done to ban "undesirable" ethnic groups from entering the country. He also opposed the teaching of foreign languages in U.S. schools, including sign language.[104]

Unquestionably, one of the most influential eugenicists and staunch supporters of U.S. immigration restriction was Madison Grant. Born in New York City in 1865, Madison Grant was a lawyer, anthropologist, author and zoologist.[105] In 1895, he founded the New York Zoological Society, spearheaded the creation of the Bronx Zoo and was an integral part of the construction of the Bronx River Parkway. Grant also initiated an effort to protect the redwood trees in Northern California. Horace M. Albright, one of the organizers of the National Park Service, once said that "no greater conservationist than Madison Grant ever lived."[106]

Grant was a strong proponent of conservation and a fierce opponent of immigration. He feverishly believed in the United States of his forefathers, and to him, an influx of people from places like southern and eastern Europe was as disastrous for the country as the loss of precious resources from deforestation or mining. Madison Grant was known for blending in with elites from the highest political and social class. He was a close friend of Theodore Roosevelt and strongly supported his third-party candidacy in 1912. He also courted wealthy donors and used these relationships to secure generous financial support for his projects, including a donation of $100,000 from the Carnegie Foundation for the Bronx Zoo, equivalent to $3.7 million in today's money.

Immigrant families arriving at Ellis Island, New York, circa 1920. *Photo courtesy of the Library of Congress.*

An Italian immigrant family arriving at Ellis Island, New York, circa 1910. *Photo courtesy of the Library of Congress.*

Madison Grant had a reputation for xenophobia and anti-Semitism. He once described the Jewish population of New York's Lower East Side as a "curse…draining off into this country [from] the great swamp" of Jewish Poland.[107] When he was tasked with proposing literacy tests for military soldiers, Grant stated that the "dwarfed and undersized Jews" were "totally unfit physically" and lacked the "moral courage for military service." He once wrote in a magazine that the United States was becoming a "dumping ground for Italians." Grant also expressed the belief that Jews and Catholics were cut from the same genetic cloth as southern and eastern Europeans and described them collectively as "half-Asiatic mongrels."

Grant lamented the high number of immigrants entering the country, which he described as the "infestation of a large and increasing number of the weak, the broken, the mentally crippled of all races drawn from the lowest stratum of the Mediterranean basin and the Balkans, together with hordes of the wretched, submerged populations of the Polish Ghetto." He claimed that "a rigid system of selection through the elimination of those who are weak or unfit, in other words, social failures—would solve the whole question in a century."[108]

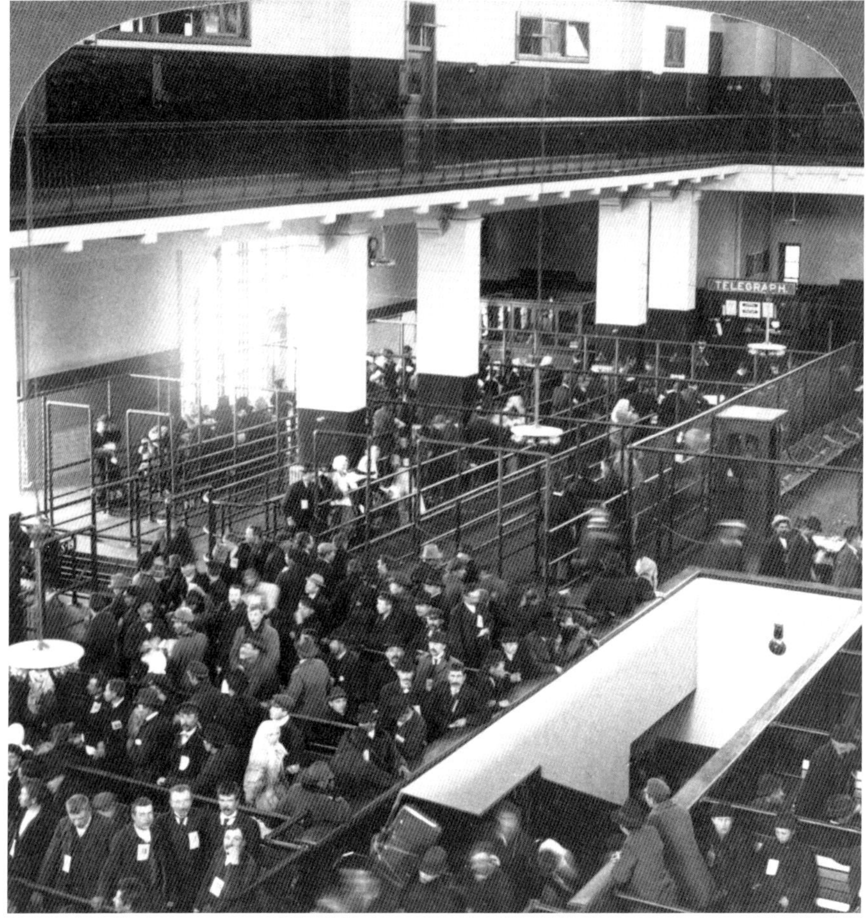

Immigrants who have just arrived from foreign countries at the Immigrant Building on Ellis Island, New York, circa 1910. *Photo courtesy of the Library of Congress.*

Madison Grant was also an author who received international acclaim for his book *The Passing of the Great Race*. Published in 1916, this virulently racist book promoted the so-called Nordic Whites as the superior race. The term *Nordic* was a purely fictional creation of Grant's that led to his fame and influence. To him, the Nordics represented the best version of the White man, a genuine "master race" facing extinction. He appears to have derived this belief from William Z. Ripley's concept of the three strong physical characteristics of Europeans (head shape, height and skin coloring), and the word *Nordic* appears to have been a substitution form Ripley's use of defining Germans as "Teutonic."

The Passing of the Great Race delivered a staunchly racist ideology and fully embraced the goals proffered by eugenicists. These included the proposal by Laughlin and others from the Eugenics Section of the American Breeders Association to engage in mass euthanasia of so-called defectives in the United States. In his book, Grant stated,

> *Mistaken regard for what are believed to be divine laws and a sentimental belief in the sanctity of human life tend to prevent both the elimination of defective infants and the sterilization of such adults as are themselves of no value to the community. The laws of nature require the obliteration of the unfit and human life is valuable only when it is of use to the community or race.*[109]

The book was so revered that it even drew great praise from Adolf Hitler, who, while in prison in 1924 for his infamous role in the Beer Hall Putsch, wrote to Grant to express adulation for the book, which he referred to as "his bible."[110]

The Passing of the Great Race was more than just a book. It was a doctrine intended to intensify the anti-immigration movement in the United States. It was widely embraced and promoted internationally by eugenicists. Historian John Higham touted Grant as "the man who put the pieces together."[111] What made Grant's book so pivotal was that it provided a direct connection between xenophobia and eugenics. This was not only seized upon by eugenicists, but it also directly fueled the anti-immigration sentiment in the United States at the time, which both compelled and inspired lawmakers in Washington, D.C., to formulate anti-immigration policies.

Madison Grant held numerous leadership roles within the Eugenics Research Association, which was established in 1913 under the auspices of the ERO at Cold Spring Harbor. Arguably the most militant arm of American eugenics, the ERA placed a sharp focus on legislative and administrative action, along with the dissemination of public propaganda about eugenics. The group boasted many of the most respected professors and intellectuals of the time, along with the nation's most radical eugenicists.[112]

OFFICIALS AT THE ERO were eager to exploit the growing anti-immigration movement for their own purposes. On February 25, 1920, Charles Davenport met with A.J. Rosanoff, chairman of the Eugenics Research Association at the Harvard Club in New York City. The main topic of discussion was immigration. Davenport offered a resolution to urge congressional legislators to establish and fund an agency to obtain family histories of each prospective immigrant seeking entry into the United States. In true eugenic fashion, the resolution stated as follows:

> *THEREFORE RESOLVED that the Eugenics Research Association approves and will urge on Congress legislation appropriate to secure the continuation for the next five years of the passport system in its application of prospective immigrants and to organize a service abroad competent to afford adequate information concerning the early personal and family history of each prospective immigrant before he leaves his own country, with the aim of admitting those of good personal and family history in their own country and rejecting those who, because of bad personal or family history, may be expected to bring with them not only undesirable personal qualities but also family tendencies of physical or mental defect or lack of self-control that will be prejudicial to the best interests of the United States.*[113]

The resolution also favored amendments to existing immigration laws that would deny "drug addicts and psychoneurotics" entry into the country and the adoption of other eugenic measures.

Since the 1880s, the United States Census Bureau had been compiling statistical data on the population it considered "defective, dependent and delinquent."[114] Those who were insane were classified as defective; the elderly and infirm were the dependent; and prisoners were the delinquent. Officials at the ERO made great efforts to expand the terminology of the census records to include the "socially inadequate," particularly along racial lines. These efforts included a campaign that ranged from mild persuasion to public castigation. Yet, citing legitimate concerns about public criticism and protest against the inclusion of the term *socially inadequate*, the Census Bureau never agreed to cooperate. It was, in fact, one of the few federal agencies that refused to partake in the eugenics movement.

The refusal of the census agency to participate in the ERO's eugenic plans proved to be a minor setback, as there was a far more agreeable and influential body willing to assist in the United States Congress. In 1920, Albert Johnson, a representative from the State of Washington, was chair of the Immigration Committee for the U.S. Congress. He was also closely aligned with other prominent Americans, like Madison Grant, who strongly supported strict immigration limits. Eager to seize upon the growing fervor, Charles Davenport sent reports to Johnson about immigrants in New York who were deemed defective. He further explained that the nation should "heavily favor the Nordics" and limit "Asiatics, Alpines, and Meds."

With the powerful support of Albert Johnson secured, it was time for Harry Laughlin to deliver. On April 16 and 17, 1920, he testified before Congress to promote the eugenic reasons for strict immigration reform. He referenced the study of the Jukes and the high costs of such defective stock. He also introduced the use and meaning of eugenic terms like *feebleminded*, *moron* and *shiftlessness* to the committee members. Laughlin's two-day testimony was published by the House of Representatives under the title "The Biological Aspects of Immigration." His presentation was so effective that it helped lead to the passage of the Emergency Immigration Restriction Act of 1921. This law led to a dramatic decrease in the total number of immigrants from Europe, which fell from approximately 650,000 in 1921 to 216,000 in 1922. The sharpest decline was in immigrants from southern and eastern Europe.[115]

Congressman Johnson was so impressed with Laughlin's presentation that he welcomed him back to provide more information to the committee and offered him the newly created title "Expert Eugenics Agent."[116] This position endowed Laughlin with the full authority to conduct racial and immigration studies on behalf of the U.S. Congress. With this powerful appointment, Harry Laughlin, who was second-in-command at the ERO, was well poised and extremely eager to take on an important role in the name of eugenics. Although he was required to provide updates to John C. Merriam, president of the Carnegie Institution of Washington, Harry Laughlin was allowed free rein to carry on his work on behalf of the congressional committee.

At the ERO, Laughlin carefully assembled all the resources available to him to craft a sensationalized presentation. Armed with congressional authority, he prepared a survey titled "Racial and Diagnostic Record of State Institutions." This survey was printed on official House of Representatives letterhead, along with his official title, "Expert Eugenics Agent," prominently displayed. He sent this survey to hundreds of hospitals, asylums and prisons in forty-eight states requesting information on those who were housed there, including their nationalities, races and the reasons for their occupancy. Notably, Laughlin only sent these forms to state-run institutions, where the population was likely poor, as opposed to the expensive private institutions more suitable to patients of wealthy families.

Laughlin also solicited the assistance of Marian Clark, chief investigator at the New York Bureau of Industries and Immigration, to prepare a survey titled "Classification Standards to Be Followed in Preparing Data for the

Schedule 'Racial and Diagnostic Records of Inmates of State Institutions.'" This survey listed sixty-five racial classifications, including "German Jew," "Polish Jew," "Spanish American" (both Indian and White), "Northern Italian," "Mountain White, "American Yankee" and others.[117] It also contained numerous types of crimes that residents of these institutions may have been involved in, including homicide, arson, disorderly conduct and so forth. All the data was collected in a massive database, analyzed and used to promote race-based immigration quotas. To round out his final report, Laughlin solicited academic assistance from the Teachers College at Columbia University, along with the Vocational Adjustment Bureau for Girls; the Life Extension Institute; and the Vineland Training School for Feeble-Minded Boys and Girls. The intent of collecting this data was clear and had long been a foundation of eugenics dogma: to show that certain racial and national types were genetically prone to crime and amoral behavior.

On October 17, 1922, Laughlin provided a brief written history of his studies and testimonial activities before Congress. He also outlined his completed work, which included "Studies in Degeneracy of Recent Immigrant Stock." Under the heading "Racial and Diagnostic Studies," Laughlin reported that he studied records of inmates at 545 state custodial institutions, where he specifically focused on ten different groups that included the feebleminded, insane, criminalistic, epileptic, inebriate, diseased, blind, deaf, crippled and dependent. He explained that the data was analyzed in comparison to the quota ratios for each class of race.[118]

On November 21, 1922, Harry Laughlin once again testified before Congress. His presentation was originally titled "Analysis of the Metal and Dross in America's Modern Melting Pot," but it was eventually published under the title *Analysis of America's Modern Melting Pot*. The voluminous report was filled with disparaging comments as Laughlin labeled the subjects he studied "human waste." Page after page, the report was rife with racial and ethnic slurs and detailed statistics regarding feeblemindedness, insanity, crime, various forms of illnesses and deformity and "all types of social inadequacy."[119] In his testimony, Laughlin crudely stated, "Particularly in the field of insanity, the statistics indicate that America, during the last few years, has been a dumping ground for the mentally unstable inhabitants of other countries."

In summarizing his study of "America's melting pot," Laughlin stated, "The logical conclusion is that the differences in institutional ratios, by races and nativity groups…represents real differences in the inborn values of the family which represent, in turn, real differences in the inborn values of the family stocks from which the particular inmates have sprung. These degeneracies and hereditary handicaps are inherent in the blood."[120] He went on to request authority to conduct separate studies of the Japanese, the Chinese, Indians, "Negroes" and Jews.

Laughlin continued to promote derogatory racial and ethnic bigotry. Under his credentials with the Carnegie Institute, he authored a treatise titled *The Parallel Case of the House Rat*, in which he likened immigrants from Europe to a rodent infestation traveling to the United States in ships.[121] Laughlin was later appointed "Special Immigration Agent to Europe." For more than six months, he traveled across Europe promoting eugenics and accumulating massive amounts of data from more than one hundred U.S. consular offices in ten countries: Sweden, Denmark, Belgium, Italy, Holland, Germany, Switzerland, England, Spain, France and the French colony of Algiers.

On March 8, 1924, Laughlin once again testified before the Congressional immigration committee. He presented a massive report titled *Europe as an Emigrant-Exporting Continent and the United States as an Immigrant-Receiving Nation*. Elaborate charts and reports were displayed promoting the link between so-called inferior races and immoral conduct. Some of the lawmakers who avidly supported draconian immigration restrictions openly read quotes from Madison Grant's *The Passing of the Great Race*. Laughlin also announced the existence of the "American Race" as "a race of white people."[122] Although he began to receive criticism for his work from U.S. congressmen, Laughlin remained undeterred and was repeatedly defended by Johnson, who stated: "Don't worry about criticism, Dr. Laughlin, you have developed a valuable research and demonstrated a most startling state of affairs."[123]

As a direct result of Laughlin's tireless efforts, which were driven by his eugenic ideals coupled with lawmakers' growing racial animus against immigrants, the House and Senate passed the Immigration Act of 1924, which was signed into law by President Calvin Coolidge on May 26, 1924. The law imposed even stricter quotas on immigrants from all non-Nordic nations. For example, the quota on immigrants from Italy was dramatically reduced from forty-two thousand per year to just four thousand. The law essentially capped new immigration from certain countries at as low as 3 percent. Laughlin's influence in the passage of this law cannot be underestimated. As the nation was considering this national immigration policy, he provided his eugenical presentation as what he called irrefutable scientific evidence to prevent the United States from succumbing to the tide of hereditary defectives from other nations. The Immigration Act of 1924 remained in full force and effect for the next forty years.

There were many forces tied to the anti-immigration sentiment in the United States. For Harry Laughlin, his work on immigration was a marked period of his career. His staunch efforts as a liaison between eugenics and U.S. immigration policy adversely affected the lives of millions of immigrants seeking entry into the United States. This included many Jews who were denied entry to the country and were later left to suffer at the hands of the Nazi regime. It is difficult to imagine many other individuals who have had such a profound effect and lasting impact on the lives of so many people than Harry Laughlin and his cohorts at the ERO, all in the name of eugenics.

PART II

THE PRACTICE

Eugenics was conceived in the early twentieth century with the wildly aspirational goal of perfecting the human race. The belief was that only those with desirable genetic traits should breed and thus produce superior offspring. Simultaneously, eugenicists grappled with how to deal with the large number of Americans they deemed to be genetically unfit to reproduce—by their estimate, as many as 15 million people. To eugenicists, these so-called defectives needed to be "eliminated from the human stock" by means of segregation, sterilization and even euthanasia.[124]

If eugenics was to truly take root, a great deal more than just scientific theory was required. The following chapters explore the planning, efforts and practical methods utilized by eugenicists to establish and promote eugenics as the necessary and true science of the age. It also explores the impact of eugenics on U.S. jurisprudence and immigration policy, along with the influence that the American eugenics program had throughout the world.

Chapter 6

THE PLOT

Resolved: that the chair appoint a committee commissioned to study and report on the best practical means for cutting off the germ-plasm of the American population.
—Eugenic Section of the American Breeders Association, 1911

From the very beginning, as he was busy securing funding and planning operations for the ERO's eugenic program, Charles Davenport was eager to formalize the organizational leadership of the ERO. The establishment of a formal corporate-like structure in accordance with all applicable laws and procedures was of paramount importance. In December 1912, he established an official Board of Scientific Directors. The initial members of the board included Alexander Graham Bell; E.E. Southard, a Harvard neuropathologist; and William Welch, a renowned pathologist from Johns Hopkins University, who served as the board's chairman.[125] Later, Thomas Hunt Morgan, a renowned geneticist from Columbia University, and Irving Fisher, a public affairs professor from Yale, also served on the board.

Davenport also helped establish affiliate groups that would operate independently but in tandem with the ERO to help promote the eugenics movement. These included the American Breeders Association, which maintained a eugenics section with its own advisory board that included some of the most prominent scientific and medical professionals of the time. Among these board members were renowned surgeon Alexis Carrol of the Rockefeller Institute for Medical Research, who would go on to win a Nobel Peace Prize for medical research; O.P. Austin, chief of the Bureau of

Members of the Board of Scientific Directors of the Eugenics Record Office, Cold Spring Harbor, New York, April 10, 1915. *From left to right*: Irving Fisher, Thomas H. Morgan and Alexander Graham Bell. *Photo courtesy of Arthur Estabrook Papers, M.E. Grenander Special Collections & Archives, University at Albany, SUNY.*

Statistics in Washington, D.C.; physiologist W.B. Cannon of Harvard; Irving Fisher from Yale; and other learned professionals. Harry Laughlin from the ERO was appointed to serve as the committee's secretary.[126]

On July 15, 1911, a meeting was convened by the board of the American Breeders Association at New York's prestigious City Club on West Forty-Fifth Street. Its agenda was to discuss the board's eugenic goals, including the best methods of dealing with the "problem of cutting off the supply of defectives," which the board estimated to be 10 percent of the American population, or as many as 15 million people. Over several subsequent meetings, the group developed a robust campaign aimed at "purging the blood of American people of the handicapping and deteriorating influences of these anti-social classes."[127]

A report, titled *Report of the Committee to Study and to Report on the Best Practical Means of Cutting of the Defective Germ-Plasm in the Human Population*, was discussed by the board in earnest. Authored by Harry Laughlin, the extensive report focused on how to deal with the "socially inadequate in the American population."[128] This ever broadening group of people included those deemed mentally deficient, criminals, the diseased and the poor. Ten groups were ultimately identified as "socially unfit" and targeted for "elimination."[129] The first was the feebleminded, a eugenics term that was previously adopted by Charles Davenport and Henry Goddard and used by other eugenicists. Although the term was never truly defined, Davenport vaguely explained that a feebleminded person was one "incapable of protecting his life against the ordinary hazards of civilization" or someone who was "deficient in some socially important trait." He ultimately conceded that feeblemindedness was difficult to properly analyze.[130] Laughlin declared feeblemindedness to be the greatest threat to U.S. society. In his report, he noted,

> *The greatest of all eugenic problems in reference to cutting off the lower levels of human society consists in devising a practicable means for eliminating hereditary feeble-mindedness. From a functional point of view, there are grades and qualities of this defect from the lowest idiot with the mentality not greater than that of a normal two-year-old child to the imbecile with the mentality not greater than that of a twelve-year-old child and the "backward" child or adult. The chronological age of such individuals is always somewhat and may be greatly in excess of their mental years.…It is the moron or high-grade feeble-minded class of individuals that constitute the greatest cacogenic menace, for these individuals, with little or no protection by a kindly social order, are able to, and do, reproduce their unworthy kind.*[131]

It was no accident that many people fell into the "feebleminded" category. It was a catch-all term used freely by all eugenicists. The remaining groups that were targeted for elimination included the pauper class; the inebriate class, or alcoholics; criminals of any kind; epileptics; the insane; the constitutionally weak; those prone to specific diseases; the deformed; and those with defective sensory organs (i.e. the blind, deaf and mute).

Laughlin's report proposed a list of dramatic remedies for dealing with the millions of Americans who were deemed socially inadequate. The first proposed remedy was life segregation (or, at a minimum, segregation during a person's peak reproductive period). Based on his data, Laughlin argued that life segregation "must, in the opinion of the committee, be the principal agent used by society in cutting off its supply of defectives."[132] Sterilization was the second proposed measure, followed by the implementation of restrictive marriage laws and customs. Public education of eugenics was fourth on the list, followed by the creation of a system of arranged mating whereby Laughlin suggested "the selection of certain potential parents, and the elimination of others." The betterment of the general environment was also proposed, along with polygamy. By far, the most frightening of the proposed remedies contemplated by the group was option number 8: euthanasia.[133] Next was a measure, dubbed Neo-Malthusianism, that proposed to limit the total number of offspring in the country. Finally, a laissez-faire approach was proposed to adopt as many eugenic goals as could be allowed. By no means were these remedies merely aspirational. Those on the American Breeders Association's board were firmly committed to implementing many if not all of these plans, and the report served as a mandate and a roadmap for the ERO and eugenicists nationwide.

At around the same time, other ERO affiliates were being established. In June 1913, the Eugenics Research Association (the ERA) was formed at Cold Spring Harbor, led by Davenport and Laughlin. The mission of this group was to initiate legislative and administrative action pertaining to eugenics, propagandize eugenic ideals of White supremacy and influence strict immigration legislation.[134] The charter members of the ERA were professional men and women in various fields, including psychology, and professors at various medical schools, including Brown, Columbia, Emory, Harvard, Johns Hopkins and Yale. Madison Grant and Lothrop Stoddard, two men widely known as race hatred fanatics, were offered leadership roles within the group. Grant gained widespread notoriety for his book *The Passing of the Great Race*, which promoted Nordic Whites as the superior race. The Harvard-educated Stoddard authored the book *The Rising Tide*

Members of the eugenic section of the American Breeders Association in 1911. Charles Davenport is seated in the front with his legs crossed. *Photo courtesy of the Truman State University, Pickler Memorial Library, Special Collections and Museums.*

Members of the Eugenics Research Association at a meeting in 1918. Harry Laughlin is kneeling in the center holding a white hat. *Photo courtesy of the Truman State University, Pickler Memorial Library, Special Collections and Museums.*

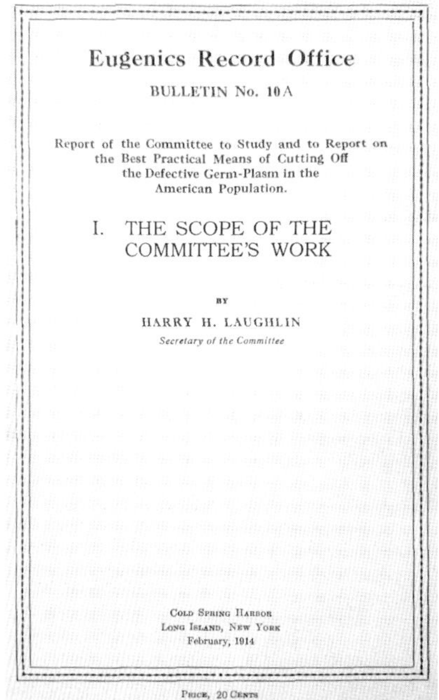

Eugenics Record Office

BULLETIN No. 10A

Report of the Committee to Study and to Report on the Best Practical Means of Cutting Off the Defective Germ-Plasm in the American Population.

I. THE SCOPE OF THE COMMITTEE'S WORK

BY

HARRY H. LAUGHLIN

Secretary of the Committee

COLD SPRING HARBOR
LONG ISLAND, NEW YORK
February, 1914

PRICE, 20 CENTS

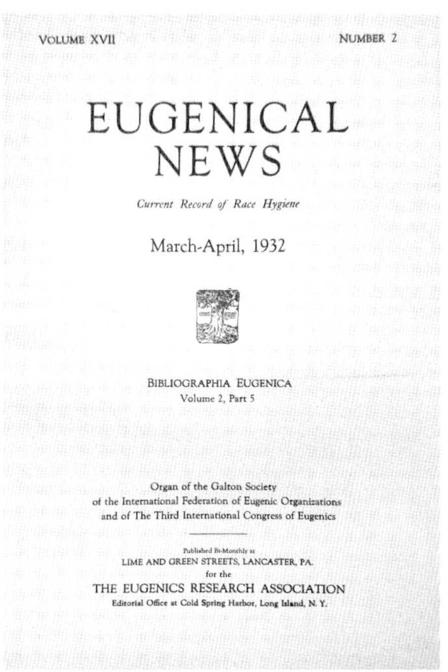

VOLUME XVII NUMBER 2

EUGENICAL NEWS

Current Record of Race Hygiene

March-April, 1932

BIBLIOGRAPHIA EUGENICA
Volume 2, Part 5

Organ of the Galton Society
of the International Federation of Eugenic Organizations
and of The Third International Congress of Eugenics

Published Bi-Monthly at
LIME AND GREEN STREETS, LANCASTER, PA.
for the
THE EUGENICS RESEARCH ASSOCIATION
Editorial Office at Cold Spring Harbor, Long Island, N. Y.

Left: *Report of the Committee to Study and to Report on the Best Practical Means of Cutting of the Defective Germ-Plasm in the Human Population*, by Harry Laughlin. Published in 1914, this report focused on ways to deal with the "socially inadequate in the American population" and targeted them for "elimination." It became both a mandate and blueprint for the American eugenics program. *Photo courtesy of Mark A. Torres.*

Right: *Eugenical News (Current Record of Race Hygiene)*, vol. 17, no. 2, March–April 1932, published for the Eugenics Research Association by the Editorial Office at Cold Spring Harbor, New York. *Photo courtesy of Mark A. Torres.*

of Color Against White World Supremacy. In it, he wrote, "You cannot make bad stock into good…any more than you can turn a cart-horse into a hunter by putting it into a fine stable, or make a mongrel into a fine dog by teaching it tricks."[135]

Any widespread movement seeking mass support requires amplification, or a proverbial mouthpiece, in order to spread its message to a multitude of people. Today, social media provides a means of such widespread and instant amplification, but in the early twentieth century, there were only newspapers and written correspondence. For the ERO, one of the main modes of mass communication was a periodic publication called the *Eugenical News*. Launched in January 1916, the *Eugenical News* had a three-man editorial board—Charles Davenport, Harry Laughlin and Morris

Steggerda—and was used to publicly share all the current news on eugenics and ongoing events sponsored by the ERO.[136] Topics covered in various editions of the *Eugenical News* included the ERO's administrative activities, important research in the field, legislative developments, academic courses and speeches that were made in furtherance of eugenics. Approximately one thousand copies of each edition were printed and shared widely by eugenics activists.[137]

The *Eugenical News* was a pivotal component of the eugenics movement. It routinely publicized the behavioral traits and mannerisms associated with certain professions from the eugenical perspective. For instance, in vol. 5 of the *Eugenical News* (1920), eugenicists discussed thalassophilia, an inherited love for the sea. In the article, the names of several captains who died or were injured in shipwrecks were published with the commentary: "Such hardy mariners do not call for our sympathy, they were following their instinct."[138] Over the years, the ERO published numerous editions of the *Eugenical News*. Several articles even included public praise for the rise of Adolf Hitler and Germany's eugenics program. The ERO also routinely published official bulletins and other eugenics-based reports that provided guidance on collecting, charting and analyzing hereditary data.

Under Davenport's leadership, the ERO greatly benefited from full funding by wealthy financiers. A board of scientific directors consisting of some of the most well-educated scientific and academic leaders of the time was empaneled. Plans were formulated to deal with the so-called defectives throughout the country. Several affiliated groups were also established, each with an escalating mission to promote and apply a series of harsh eugenic measures throughout the United States and the entire world. Periodic publications like the *Eugenical News* amplified the eugenics movement to a mass audience. Collectively, these elements provided the foundation of the U.S. eugenics program, and a nationwide practice of eugenics was ready to be launched.

Chapter 7

THE PEDIGREE HUNTERS

Certainly, one of the sins of the State in handling its problem of the feebleminded should be to dry up the springs from which they arise.
—Charles Davenport, 1939

The mere thought of eugenical experiments understandably conjures dark fears of the barbaric torture of human beings. The Nazi regime certainly engaged in unthinkable acts of atrocity under the guise of eugenic science. These acts, which will be discussed in a later chapter, were directly inspired by the American eugenics program. However, torture and murder were but the ultimate stages of the eugenic crusade. After all, even the most diabolical movements have their origins, and for eugenicists, it began with forethought, planning and data collection.

A large part of the work conducted by Charles Davenport and his ilk at the ERO can ultimately be described as the collection and study of family pedigrees. The data that was gathered en masse was necessary to buttress the entire pseudoscience and establish a propaganda machine aimed at cementing a global narrative about the permanent need for eugenics. This information was required to validate eugenics as a science in support of its foregone conclusion that society was overburdened with a multitude of genetically inferior human beings. The work of information gathering should not be viewed as harmless or perfunctory. To the contrary, the ERO's work of gathering hereditary data was the singular most important element of the eugenics movement in the United States and throughout the world. Data collection was the very foundation for the practice of eugenics, and since it enabled the barbaric acts that would follow, it should be viewed as equally sinister.

As far back as the seventeenth century, scientists were utilizing what was known as the scientific method for their research. In essence, a theory or hypothesis arose. Scientists would then ask questions and conduct experiments to either prove or disprove the proposed hypothesis.[139] Eugenicists practiced this in reverse. Having already formulated a preconceived notion of heredity, typically premised on classist and racist beliefs, they would then seek to amass the data to fortify this theory.

The constraints of time hindered eugenicists. Genetic scientists, like Thomas Hunt Morgan, who studied the genetic traits of fruit flies at Columbia University in New York, had the benefit of time. Able to breed just ten days after hatching, fruit flies enabled his team to make and study predictions of inheritable traits at a rapid pace. The same was true for those who studied the genetic traits of crops. However, eugenicists aimed to study the hereditary traits of human beings, who have a far greater lifespan than fruit flies or crops. To offset this lengthy period of time, eugenicists made assumptions in the study of human heredity, and in doing so, they relied entirely on the data that was gathered by researchers like those at the ERO. Since these assumptions were premised on racist and classist beliefs instead of true scientific research, the data they gathered was overwhelmingly unreliable and most often fabricated. Later, Thomas Hunt Morgan began to understand that studying human genetics was far more complex than his work with fruit flies or Mendel's work with peas. As a result, he began to doubt the simplicity of the work that eugenicists relied on and touted. Seeking to distance himself from this flawed science, he quietly resigned from serving on the ERO board and asked that his name be removed from all future ERO publications.

The American eugenics program was as lofty as it was sinister. It was premised on the need to collect hereditary data from subjects all over the country and even the world. Davenport was inspired by the work of Gregor Mendel, and he fully embraced the notion that evolution works for human beings in the same way it does for animals or plants. He believed that high incidences of certain characteristics within family pedigree charts conclusively proved that those characteristics were inheritable—and that some of them needed to be eliminated from society. He argued that patterns of heritability existed in insanity, epilepsy, alcoholism, pauperism, criminality

This page: The Eugenics Record Office stored millions of index cards, pedigree charts, photographs and other data for eugenic research in specialized fireproof file cabinets at the Cold Spring Harbor facility. *Photos courtesy of the Truman State University, Pickler Memorial Library, Special Collections and Museums.*

Pedigree charts and other records were stored in these fireproof file cabinets at the Eugenics Record Office in Cold Spring Harbor, New York. *Photo courtesy of Cold Spring Harbor Laboratory Archives, New York.*

and feeblemindedness.[140] As a result, the ERO meticulously collected massive amounts of information on its subjects. This included millions of documents, notes, letters, index cards, pedigree charts and photographs and other data deemed necessary for eugenics research, all of which was stored in specialized fireproof file cabinets at the Cold Spring Harbor facility.

The widely ambitious program was premised on three main phases. The first was to identify and target areas throughout the country where people possessed eugenically undesirable traits. The ERO estimated this group made up as much as 10 percent of the American population, or the equivalent of 15 million people. They nicknamed this group the "submerged tenth," and their hunt for human subjects was relentless.[141] The second phase of the program was to dispatch numerous field agents to visit these subjects to interview them and make observations. Finally, all this information was collected and catalogued at the ERO and later used to amplify its eugenic ideals through a robust public relations campaign. In three decades of operation, ERO officials perfected this three-step process.

This page: Workers reviewing records at the Eugenics Record Office in Cold Spring Harbor, New York. *Photos courtesy of Cold Spring Harbor Laboratory Archives, New York.*

THE CREDENTIALS OF CHARLES Davenport were beyond question. He was well educated and popular in the burgeoning field of hereditary science. He was both a practitioner and a professor who had published numerous books and medical treatises on the topic and lectured in many classes. Most importantly, he was director of a thoroughly planned and well-funded eugenic program at Cold Spring Harbor. These credentials afforded Davenport the luxury of attaining unfettered access to subjects for his research. He keenly understood that studying large captive groups of people contained within controlled environments was an efficient way to amass large amounts of eugenically biased data at a rapid pace. Davenport often wrote to Mary Harriman touting the access he was routinely given to numerous facilities that housed large numbers of subjects, along with their familial records. These facilities included numerous prisons, psychiatric institutions, hospitals, orphanages and poorhouses. Davenport even studied draftees of the U.S. military. ERO officials also visited Native American reservations and rural areas with people who had been marginalized and were destitute and often uneducated.

The archives at Cold Spring Harbor, the American Philosophical Institute and other similar facilities are replete with large collections of correspondence, both professional and personal, between Davenport and those directly in control of these types of facilities, including Henry Goddard of the Vineland Training School for Feeble-Minded Boys and Girls; Katharine Bement Davis, who directed the New York Parole Commission; Dr. Edward Humphries at the Letchworth Village Psychiatric Hospital at Stony Point, New York; and officials at other similar locations. In one letter, Davenport blithely wrote to the director of the American Museum of Natural History to request skulls of Long Island's Shinnecock Indians from their collection for his eugenic studies. It was abundantly clear that Davenport's credentials enabled ERO officials to have unfettered access to a vast number of subjects in a captive audience setting to conduct eugenic research.

IT WAS NO SMALL task for the relatively small staff at the ERO to visit and create family records at a multitude of locations throughout New York and the nation. Like-minded assistants had to be recruited to collect the necessary data. To accomplish this, the ERO developed a robust training program at Cold Spring Harbor for recruits to learn the necessary skills for this task. The primary source of training material was Davenport's *Trait Book*.[142] The purpose of the *Trait Book*, published in 1912, was to amass a list of inheritable human traits and to standardize terms to be used by eugenic research agents

Training class at the Eugenics Record Office, circa 1915. Between 1911 and 1924, an estimated 250 field-workers were trained and dispatched by the ERO to conduct eugenics research across the nation. The researchers were paid a sum of seventy-five dollars per month and had to commit to a minimum of one year of work in the field. *Photo courtesy of the Truman State University, Pickler Memorial Library, Special Collections and Museums.*

in the field. In the book, Davenport assigned numerical codes to a wide spectrum of human characteristics. In doing so, he erected ten fundamental classes of traits, which were arranged with numerous subdivisions. He focused on various biological systems within the human body that included the circulatory, excretory, muscular, nervous, reproductive, respiratory and skeletal systems, as well as the sense organs. He also included general and mental traits, criminality, diseases and occupations as specified categories.[143]

The ERO had little difficulty in attracting a small army of qualified individuals for the program. Most of the trainees were well-educated women who had recently graduated from esteemed institutions such as Cornell, Harvard, Radcliffe, Vassar and Wellesley, along with other reputable colleges.[144] Between 1911 and 1924, an estimated 250 field-workers were dispatched by the ERO to conduct studies on numerous subjects nationwide.[145] They were paid a sum of seventy-five dollars per month and had to commit to a minimum of one year of work in the field.

These field researchers visited numerous facilities and rural locales throughout the country to interview and observe patients or groups of

Top: Charles Davenport lecturing to trainees at the Eugenics Record Office, circa 1918. *Photo courtesy of the Truman State University, Pickler Memorial Library, Special Collections and Museums.*

Bottom: Training class at the Eugenics Record Office, circa 1918. Harry Laughlin is standing near the blackboard. *Photo courtesy of the Truman State University, Pickler Memorial Library, Special Collections and Museums.*

people. Tabulated summaries of the field-workers' research were created. They depicted how each researcher was assigned to study various conditions, including insanity, epilepsy, feeblemindedness and albinism, as well as topics like skin color, polygamous families and sterilization.[146]

ERO researchers collected anthropometric measurements of the bodies of those who were examined, which included measurements of the skulls, along with their height and weight. Hair samples were also collected. Researchers probed family histories, and pedigree charts were created. Since the subjects were asked to recall and give their impressions of family members, both past

Training class at the Eugenics Record Office, circa 1919. Charles Davenport is standing in the top row to the far right. *Photo courtesy of the Truman State University, Pickler Memorial Library, Special Collections and Museums.*

Training class at the Eugenics Record Office, circa 1922. Harry Laughlin is standing in the top row to the far right. *Photo courtesy of the Truman State University, Pickler Memorial Library, Special Collections and Museums.*

"The Investigators" conducting research in Amherst, Virginia. Eugenics field officer Ivan E. McDougle is kneeling in the rear. In the front, from left to right, are field assistants Martha Gwendolyn Watson, Martha Lobingier and Eleanor Harned. *Photo courtesy of Arthur Estabrook Papers, M.E. Grenander Special Collections & Archives, University at Albany, SUNY.*

Harry Laughlin displaying a mask during a training. *Photo courtesy of the Truman State University, Pickler Memorial Library, Special Collections and Museums.*

and present, such data was inherently unreliable and probabilistic at best. Nevertheless, the information was never questioned and promptly secured and relied on by eugenicists as entirely factual.

ARMED WITH UNFETTERED ACCESS and a well-trained cadre of devoted field assistants, it was time for the ERO's work to begin—work which can most aptly be described as a relentless hunt for human subjects. Some of the

Two trainees seated at the tent colony in the sheep pasture at Cold Spring Harbor, New York. *Photo courtesy of the Truman State University, Pickler Memorial Library, Special Collections and Museums.*

A eugenics field-worker training conference, circa 1912. *Photo courtesy of Cold Spring Harbor Laboratory Archives, New York.*

Eugenics field-worker training at Kings Park, New York. *Photo courtesy of Cold Spring Harbor Laboratory Archives, New York.*

subjects that ERO field agents traveled to study were inmates at Sing Sing Prison in Ossining, New York; patients at Letchworth Village Psychiatric Center in Rockland County; albino families in Millerton, New York; insane patients at the New Jersey State Hospital in Matawan, New Jersey; and Amish families in Pennsylvania.[147] On Long Island, New York, field researchers visited the Unkechaug and Shinnecock reservations and the Kings Park Psychiatric Center. They also studied circus performers and their families at the so-called freak shows at Coney Island in Brooklyn. By as early as 1913, a multitude of data had been retrieved, indexed and stored in specialized fireproof cabinets at the ERO in Cold Spring Harbor.

Comprehensive pedigree charts were created with familial information retrieved by the field-workers and catalogued based on a wide range of real or perceived mental or physical familial defects. Some of these charts were titled "4 Generations, One Almshouse, at One Time,"[148] "Hovel Type of Source of Defectives,"[149] "Heredity Predisposition to Mental Illness"[150] and "Poorhouse Type of Source of Defectives: First Study of Epilepsy.[151]" Elaborate drawings, charts and graphs were also added to supplement the field-workers' training, including a "Tree of Life" drawing with an explanation of eugenics[152] and an "Abacus for Illustrating of the Human Germ Plasm."[153]

One regular target for eugenicists worldwide was almshouses, which were often referred to as poorhouses. A captive population of paupers and infirm patients was a suitable source for eugenicists to collect eugenic data in large numbers. In the late nineteenth century, the New York State Board of Charities designated five facilities to provide care for the state's growing indigent population.[154] On Long Island, in Suffolk County, the Yaphank Alms House opened in 1874. The facility was used to house hundreds of indigent men, women and children from Long Island and throughout the state.

Jonathan Baker (1853–1923) served as the superintendent of the Yaphank Alms House. In the summer of 1913, Baker chaired a meeting of the State Association of Superintendents in Binghamton, New York. During a speech, he publicly declared his support for eugenics, stating, "In every community, there exist more or less persons who in the ordinary course of life cannot be made good citizens and to this class every effort should be made to prevent

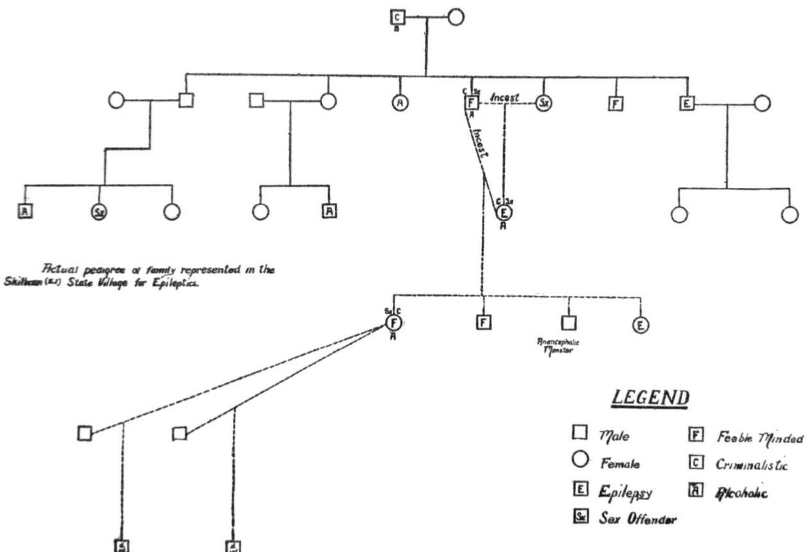

This page and opposite: Pedigree charts purporting to depict the origins of certain mental disorders. *Photos courtesy of Mark A. Torres.*

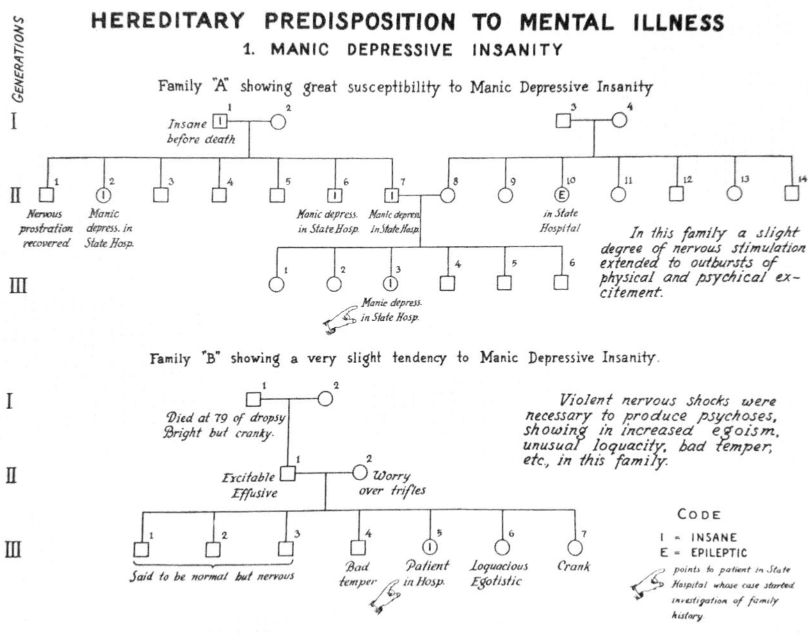
GENERATIONS
HEREDITARY PREDISPOSITION TO MENTAL ILLNESS
1. MANIC DEPRESSIVE INSANITY
Family "A" showing great susceptibility to Manic Depressive Insanity
Insane before death
Nervous prostration recovered
Manic depress. in State Hosp.
Manic depress. in State Hosp.
Manic depress. in State Hosp.
in State Hospital
In this family a slight degree of nervous stimulation extended to outbursts of physical and psychical excitement.
Manic depress. in State Hosp.
Family "B" showing a very slight tendency to Manic Depressive Insanity.
Died at 79 of dropsy Bright but cranky.
Violent nervous shocks were necessary to produce psychoses, showing in increased egoism, unusual loquacity, bad temper, etc., in this family.
Excitable Effusive
Worry over trifles
Said to be normal but nervous
Bad temper
Patient in Hosp.
Loquacious Egotistic
Crank
CODE
I = INSANE
E = EPILEPTIC
points to patient in State Hospital whose case started investigation of family history

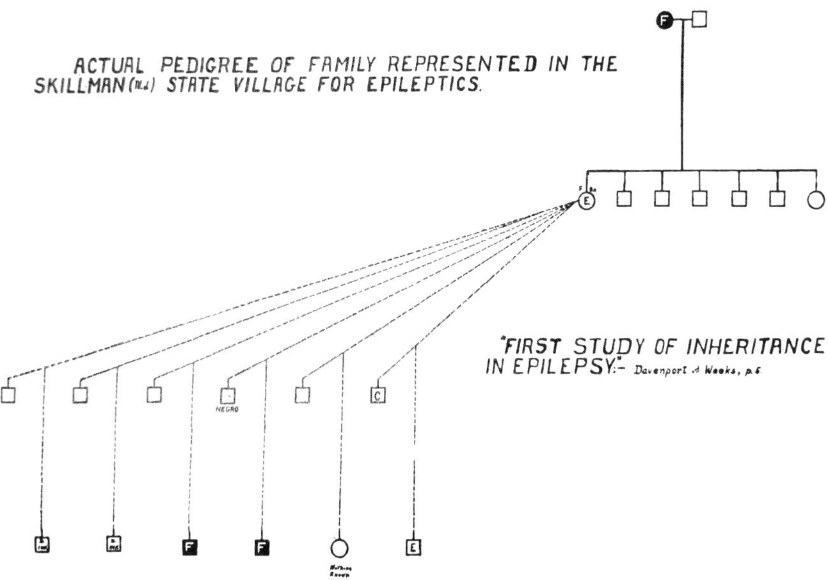
POORHOUSE TYPE OF SOURCE OF DEFECTIVES.
ACTUAL PEDIGREE OF FAMILY REPRESENTED IN THE SKILLMAN (N.J.) STATE VILLAGE FOR EPILEPTICS.
"FIRST STUDY OF INHERITANCE IN EPILEPSY." – Davenport & Weeks, p.6
NEGRO

Four Generations in One Almshouse at One Time, at Yaphank, Suffolk Co; N.Y.–July 27, 1917.

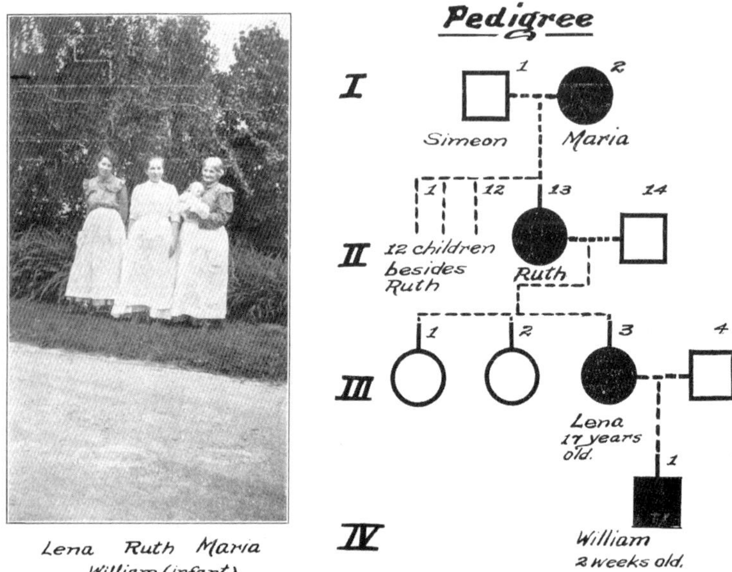

A border-line family, in which illegitimacy runs high, not quite able to care for itself in organized society.

A eugenic pedigree chart titled "Four Generations in One Almshouse at One Time, at Yaphank, Suffolk Co., N.Y.—July 27, 1917." *Photo courtesy of Cold Spring Harbor Laboratory Archives, New York.*

their propagation."[155] Using a common argument made by eugenicists of the time, Baker cited the need to reduce or eliminate altogether the cost of treatment for the insane, the feebleminded, the blind and the deaf, as well as prisoners and residents in almshouses throughout the state.

ERO officials were freely granted access to conduct eugenical studies at the Yaphank Alms House. On July 27, 1917, ERO researchers who visited the facility compiled a study of one particular family. A pedigree chart titled "Four Generations in One Almshouse at One Time at Yaphank, New York" was created, which was accompanied by a photograph of three women of the same family who resided at the almshouse, one of whom was holding an infant child.[156] A description at the bottom of the pedigree card states, "A border-line family, in which illegitimacy runs high, not quite able to care for itself in organized society."

Although this study included four generations of family members who were deemed genetically unfit, it fully ignored any of the socioeconomical, nongenetic factors that may have contributed to the despair of this family. Nevertheless, it was used in support of the eugenically driven principle that people in this condition should not be allowed to procreate. Just ten years later, a similar argument was made about three generations of the same family in Virginia before the U.S. Supreme Court in the infamous *Buck v. Bell* case, during which Justice Oliver Wendell Holmes Jr. stated, "Three generations of imbeciles are enough."

Native American reservations on Long Island were also targeted by eugenicists. In 1913, Charles Davenport secured access to the Shinnecock Reservation and returned there periodically over two decades. In 1932, Davenport and Morris Steggerda compiled a report titled *The Shinnecock Indians of Long Island, New York*.[157] Although unpublished, the voluminous report details the history of the Shinnecock Indians, amassed from various publications, and includes numerous photographs of Shinnecock families, the homes they resided in and the land they inhabited. A wide array of pedigree charts was presented, alongside anthropometric measurements and other physical descriptions taken by ERO field-workers who assisted in the project.

In the summer of 1923, several ERO trainees were dispatched to the Unkechaug Reservation on Poospatuck Creek, near Mastic, New York. In an attempt to put their subjects at ease, the trainees falsely claimed that the purpose of the study was to determine if their bloodlines were pure. They also brought food and other gifts to gain their subjects' trust.[158] The four-day study took place between July 20 and 23 and was primarily used for training purposes. Three adults and twelve children, some of whom lived both on and away from the reservation, agreed to participate in the study. Charles Davenport instructed his trainees to use administer Terman's Standard Revision of the Binet-Simon Intelligence (IQ) Test during the project.[159] Anthropometric measurements of several individuals were taken, and hair samples were collected. These measurements were compared to so-called racial norms, which analyzed head and nose breadth, stature, hair form and lip thickness. A full report with nearly one thousand pages of documents and pictures was compiled.

Although the studies and reports conducted by the ERO on the Native peoples of Long Island were never published, they were used for training purposes. Field-workers were given ample time to hone their skills before fanning out across the country for eugenic studies and data collection. Furthermore, it is believed that these studies were relied on to make decisions that had adverse effects on Native peoples. John A. Strong, professor emeritus of history at Long Island University and author of numerous books on the history and culture of Native Americans, has argued that the eugenically biased data derived from these studies was used by the Bureau of Indian Affairs in the Indian Reorganization Act to the detriment of the Native population.[160]

According to Strong, the ERO's report displays a lack of professional standards and is rife with racial and class bias. For instance, one trainee named Martha Abbott referred to the homes of those they studied as "shacks" and stated that "the inmates of the house were not overly clean."[161] Abbott echoed Davenport's expectation that "mixed races" would score poorly on IQ tests. When one child scored above average on the test, she arbitrarily noted that "114 seems a rather high I.Q. for a child of [name deleted]'s probable pedigree."

Similar eugenic studies were conducted in local communities on Long Island and throughout New York State. In August 1921, a group of neighboring families on Babylon Road in Commack were subjected to eugenic studies.[162] A similar study was conducted in July 1924.[163] Harry Laughlin personally assigned Arthur Estabrook to "search for degenerates in the isolated valleys around the upper Hudson River," which yielded scores of pedigree charts.[164] Estabrook was well versed in the program, as he previously conducted eugenic studies with Charles Davenport on the island of Jamaica. Estabrook conducted similar studies on families in Ulster County, New York, and central Pennsylvania.[165]

Another prime target for eugenics research was prisons, and for this, eugenicists benefited greatly from a relationship with Katharine Bement Davis (1860–1935). A graduate of Vassar, Davis became the first female parole commissioner of the New York State Parole System and served as the executive secretary of the Bureau of Social Hygiene Correction Commissioner in 1914.[166] Prior to her appointment to this powerful position, Davis spent thirteen years as superintendent of the Reformatory for Women at Bedford, New York, where she conducted numerous studies of the criminal tendencies of female prisoners. Davis was a staunch supporter of eugenics who, in her role as commissioner,

Katharine Bement Davis was a strong supporter of eugenics and maintained close ties with Charles Davenport. *Photo courtesy of the Library of Congress.*

openly welcomed the scientific study of "the defective and delinquent classes" within state-run institutions. She aggressively urged the use of "our great institutions as human laboratories."[167]

Davis was also instrumental in assisting Harry Laughlin with the release of his seminal book *Eugenical Sterilization in the United States*. While many publishers were skeptical of taking part in a project that openly promoted compulsory sterilization, Davis eagerly served as a liaison between the ERO and several prospective publishers. In one letter, dated June 2, 1920, Laughlin himself asked Davis to forward a copy of the manuscript to Macmillan. The publisher reviewed the manuscript and eventually rejected the project, largely due to high production costs and its structural composition. In a subsequent letter, dated February 2, 1921, Davis began to express hesitation about a book that had to do with "direct propaganda favoring sterilization legislation." However, she remained a staunch advocate for eugenics and made all best efforts to locate a publisher for the project. Ultimately, her powerful position overseeing the New York State Parole Commission made Katherine Davis a staunch ally of the ERO and the eugenics movement.

Support for eugenics research was also garnered by administrators at the Clinton Correctional Facility in Dannemora, New York, and the Auburn State Prison in Auburn, New York.

PSYCHIATRIC INSTITUTIONS IN NEW York and throughout the nation were perhaps the ideal setting for eugenic research. One such facility was the Kings Park Psychiatric Center, which was constructed on over eight hundred acres on the shores of the Long Island Sound in Suffolk County, New York, in 1885.[168] The facility was built to alleviate overcrowding at the Kings County Lunatic Asylum in Brooklyn, New York. In 1895, New York State assumed full control of the asylum.

By the turn of the century, Kings Park housed approximately 2,700 patients and nearly 500 staff members. By the 1930s, more buildings were constructed to deal with the growing population of patients, and by 1954,

Pictured in the center is the Clinton Prison in Dannemora, New York, circa 1915. Originally built in 1845 on two hundred acres of land, the facility was first used to house laborers from a nearby iron mine before becoming a prison. The facility was nicknamed Little Siberia due to its harsh winters and remote location. *Photo courtesy of Arthur Estabrook Papers, M.E. Grenander Special Collections & Archives, University at Albany, SUNY.*

the hospital had approximately 9,300 patients on a sprawling property that boasted over 150 buildings. Kings Park was fully self-sufficient, with a power plant, maintenance buildings, livestock for food, dormitories for staff and even its own railroad station.

Through the 1950s, Kings Park created a great deal of controversy when it adopted aggressive forms of treatment that included prefrontal lobotomies and electroshock therapy. The buildings were later described as unhygienic, and the quality of food was poor. Over the years, scientists developed new medications, including the antipsychotic Thorazine, that allowed an increasing number of patients to be treated at home. As the number of patients at the asylum continued to decline, a growing number of buildings at the facility were demolished, and by 1996, Kings Park had closed its doors for good.

A massive facility housing thousands of patients in a self-contained environment coupled with administrators who staunchly supported eugenics was the optimal setting for ERO officials to conduct research. In 1914, Dr. William Austin Macy, the medical superintendent of Kings Park, publicly shared the eugenic measures being utilized at this facility. "We are carrying out our plans," he stated, "on the lines of heredity and have made careful genealogical researches into numerous cases, and the results conclusively prove that the transmission of insanity from one generation to another is an assured fact."[169]

Dr. Macy asserted that eugenics was the only measure that could prevent insanity, and while he explained that the work at Kings Park was initially aimed at tracing familial lines, he mentioned that sterilization would be an inevitable and necessary practical measure to solve the problem of insanity. He added that the facility was "preparing useful data and information for the benefit and guidance for future generations for the successful prevention of insanity." Dr. Macy explained that the administration of Kings Park was in "frequent consultation" with Charles Davenport and that he was "heartily in sympathy with the science of eugenics."

Other psychiatric institutions that openly facilitated eugenic research throughout Long Island included the Central Islip State Hospital and the Pilgrim State Hospital in Brentwood. Hospitals in the upstate region of New York that also enthusiastically supported eugenics included the Dannemora State Hospital, the Gowanda State Hospital, the Rome State Custodial Asylum and the State Hospital in Buffalo, New York.

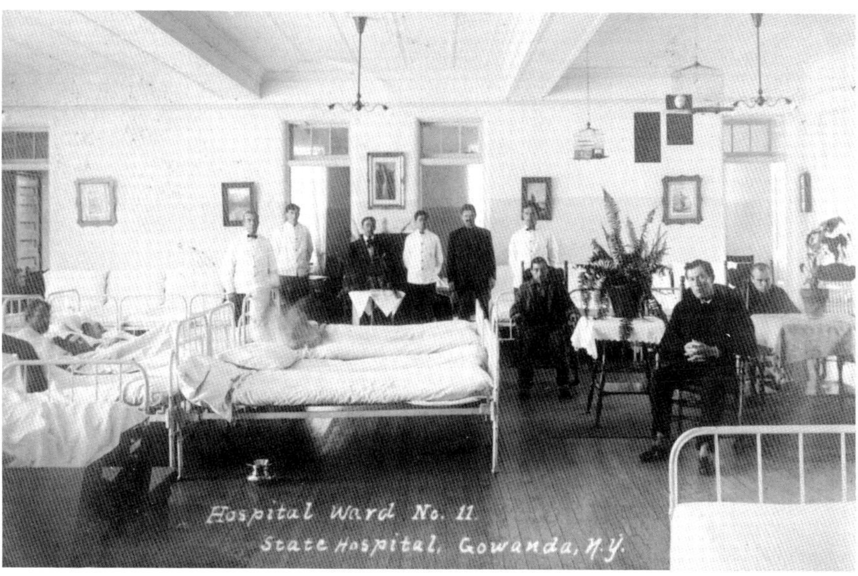

Patients and attendants at Ward No. 11 of the Gowanda State Hospital in Collins, New York. *Photo courtesy of Robert Lewis and AsylumPostcards.com.*

The State Hospital, Kings Park, L. I.

This page: The Kings Park Psychiatric Center in Kings Park, New York, February 8, 1906. This facility was in operation from 1895 to 1996. At its peak, Kings Park treated 9,300 patients and maintained over 150 buildings. The facility was fully self-sufficient with its own power plant, maintenance buildings, livestock for food, dormitories for staff and even its own railroad station. *Photos courtesy of Robert Lewis and AsylumPostcards.com.*

This page: Dannemora State Hospital in Dannemora, New York. *Photos courtesy of Robert Lewis and AsylumPostcards.com.*

This page and opposite: The state hospital in Buffalo, New York. *Photos courtesy of Robert Lewis and AsylumPostcards.com.*

Approximately eighty miles north of the former Kings Park facility is Thiells, New York. This small hamlet in Rockland County was the home of another large psychiatric hospital called Letchworth Village, which opened in 1911, a residential institution used to house physically and mentally disabled patients of all ages.[170] The village contained 130 mostly two-story buildings, spread out on a sprawling property of more than two thousand square acres. At its peak, the facility housed nearly four thousand patients and, over the years, drew sharp criticism after the public learned of the neglectful and even harmful treatment of its patients. Letchworth Village was in operation until its closure in 1996.

As he had done at similar facilities, Charles Davenport developed a strong working relationship with various administrators at Letchworth Village. He regularly corresponded with Charles S. Little, the superintendent at Letchworth from 1913 to 1926. He was also in close contact with Edward Humphreys, director of research at Letchworth Village, who readily provided Davenport with lists of all patients along with their medical history and family data for further study and exploitation by the ERO.[171] Information on pregnancies of active and discharged patients was also provided.

During the summer of 1929, Charles Davenport collaborated with doctors at Letchworth Village to perform the castration of a thirteen-year-old male patient at Letchworth Village. The procedure was purely for eugenic purposes, as he wanted to retrieve tissue and analyze the boy's chromosomes in hopes of learning the causes of mental retardation.[172]

PILGRIM STATE HOSPITAL, BUILDING NO. 25, BRENTWOOD, LONG ISLAND, N. Y.

Building no. 25 of the Pilgrim State Hospital in Brentwood, New York. *Photo courtesy of Robert Lewis and AsylumPostcards.com.*

The Women's Building of the Gowanda State Hospital in Collins, New York. From the enactment of New York's eugenic sterilization law in 1912 until its repeal in 1920, most of the state's sterilization procedures were performed on females at this facility. *Photo courtesy of Robert Lewis and AsylumPostcards.com.*

The Rome Custodial Asylum in Rome, New York. On September 17, 1915, the board of examiners at the facility filed for litigation to have a twenty-two-year-old patient sterilized against his will pursuant to New York's eugenic sterilization law. *Photo courtesy of Robert Lewis and AsylumPostcards.com.*

In the nineteenth century, a British physician named Langdon Down described a medical condition that was previously called "Mongolism." Now called Down syndrome, the condition occurs in people who are born with three (instead of the ordinary two) of the twenty-first chromosome.[173] For several years, Charles Davenport had intensely studied and written about this condition, and he openly advocated for aggressive research on patients with Down syndrome housed in state-run institutions.

Davenport identified a thirteen-year-old boy from Letchworth Village on whom to perform the experimental dissection of one of his testicles for the purposes of studying its tissue. He was fascinated with the "possibilities of making a study of the chromosomal conditions in cell divisions of the testes of a Mongoloid." Davenport selected the boy for this procedure for two reasons. The first was based on his review of the child's medical records—a clear byproduct of having unfettered access to all medical records at the facility. The other reason was based on Davenport's speculation that the boy exhibited a "marked eroticism," which, Davenport believed, likely bothered the boy.[174]

Davenport assembled a medical team to perform the surgery, including George Washington Corner, an anatomist who received his medical degree from Johns Hopkins University. Even though Corner was not a licensed surgeon in New York and had only, up to that point, performed surgeries on dogs, he was more than eager to assist. Davenport trusted his work, and according to Corner, Davenport had once provided him with "a pair of human ovaries" for a study on twins. The ovaries were taken from a woman who had died a short time after delivering triplets in a New York hospital.[175] Davenport also recruited Theophilus Painter, who earned his PhD from Yale University. Both men were much younger than Davenport, but they shared his passion for eugenics, and each had spent time studying eugenics at Cold Spring Harbor.

Weary of adverse public criticism, the team discussed the need to obtain parental consent to conduct the procedure. However, the boy's father was already deceased, and his mother was described as someone with "low mentality." Thus, obtaining legal consent was challenging. However, these concerns were resolved when the clinical director of Letchworth Village informed Davenport that he had obtained written and signed consent from the boy's mother. An explanation was neither requested nor provided of how this consent was obtained, but given the woman's allegedly limited mental capacity, it seems highly unlikely that she would have thoroughly understood the nature of the procedure that was to be performed on her son.

The procedure took place in August 1929. Corner performed the surgery, and Painter was tasked with providing a report on his analysis of the tissue samples from the castrated testicle to search for abnormalities. No known records exist to suggest that Davenport was able to conclusively prove his theories on the causes of mental retardation and the study ultimately revealed no new insights.

Davenport's strong working relationship with Letchworth Village continued for many years. In a letter dated March 25, 1935, he wrote to Dr. Edward Humphries at Letchworth Village to request photographs of the skull of a patient whom he had been studying.[176] He also requested a complete list of all patients, including those with Down syndrome, whom he called "Mongols," and scheduled future visits with Humphries at that facility.

In a letter dated July 26, 1939, Davenport wrote to Elizabeth Buck, a special investigator at Letchworth Village, thanking her for sending him the institution's card catalogue of present and former patients' pregnancies.[177] In the letter urging the publication of the data, he wrote,

Undated photo of Charles Davenport working at his desk. *Photo courtesy of the Truman State University, Pickler Memorial Library, Special Collections and Museums.*

> *Certainly, one of the sins of the State in handling its problem of the feebleminded should be to dry up the springs from which they arise. If the data were published, as they should be, they will bring home clearly to the legislature the fact that the work in the direction of drying up the springs has been very imperfect, that a great advance would be secured were it possible to sterilize the girls before they are discharged from the Institution.*

This request would never be finalized, as the ERO would close a mere five months after the date of that letter. Nevertheless, the level of cooperation that Davenport established with the superintendents, doctors and other authoritative figures within Letchworth Village appeared to be limitless.

During the First World War, the National Academy of Sciences established the National Research Council, which was led by a group of psychologists including Henry Goddard and Robert Yerkes. Beginning in May 1917, the group conducted intelligence tests on men who were drafted by the U.S. Army. As Yerkes noted, the testing was "not primarily for the exclusion of intellectual defectives, but rather for the classification of men in order that they may be properly placed in the military service."[178]

Using a variation of Goddard's Binet-Simon test, recruits at Camp Devens in Massachusetts, Camp Dix in New Jersey, Camp Taylor in Kentucky and Camp Lee in Virginia were all tested. The "Alpha" test was administered to English-speaking draftees, and the "Beta" test was for non–English speaking draftees. The tests were biased in favor of scholastic aptitude and entirely dependent on the educational and cultural background of the person being tested. This was problematic for many of those tested, who were barely educated and were raised by poor families in rural settings. For example, one question on the Alpha test required knowledge of specialized automobiles.[179] The knowledge needed to answer this and other questions was well beyond what the average U.S. citizen was able to attain. As one examiner noted, "It was touching to see the intense effort…put into answering the questions, often by men who never before had held a pencil in their hands."[180]

When the testing was completed and the results tabulated, it was determined more than half of the 1.7 million draftees were eugenically diagnosed as "morons." Eugenicists seized on these results and argued that the nation's military, along with U.S. society overall, was in desperate need of widespread eugenic measures.

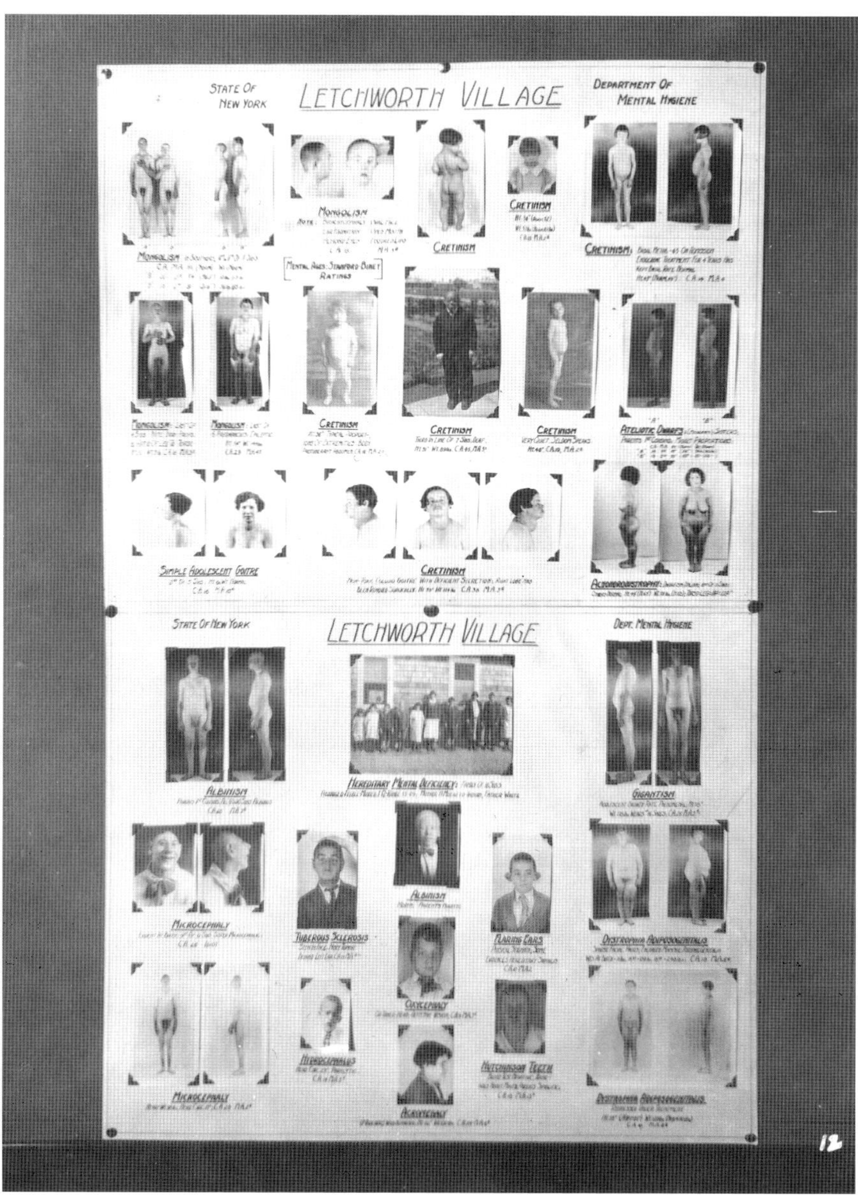

A eugenics exhibit of patients at the Letchworth Village Psychiatric Center in Thiells, New York. This information was on public display at the Third International Eugenics Congress in 1932. *Photo courtesy of the Truman State University, Pickler Memorial Library, Special Collections and Museums.*

This page and opposite: Some of the buildings used at Letchworth Village remain as a stark reminder of this dark history. *Photos courtesy of Mark A. Torres.*

Above: The Old Letchworth Village Cemetery (1914–1967) near the grounds of the former facility, where many of the patients are laid to rest in marked, numbered plots. *Photo courtesy of Mark A. Torres.*

Right: A memorial plaque titled "Those Who Shall Not Be Forgotten" at the Old Letchworth Village Cemetery honoring those who died at Letchworth Village. *Photo courtesy of Mark A. Torres.*

During the latter part of the nineteenth century, Coney Island's Manhattan and West Brighton Beaches became popular destinations for middle-class Americans. Along with access to the Atlantic Ocean, there were many types of recreational activities available in the area and a wide variety of performances to entertain visitors. This included the wildly popular circus performers at the so-called freak shows.[181] Naturally, eugenicists believed that the abnormalities of these performers were prime examples of degenerate heredity, and the glorification of these performers was deeply troubling to them. Thus, as Davenport had done with patients in mental asylums, almshouses and prisons, Davenport and the ERO embarked on a project to study the people of this isolated community. Their goal was to identify and use these performers as model examples of the degenerative traits that needed to be eliminated from society.[182]

There was a wide variety of performers at Coney Island, including Lionel, the lion-faced boy. Promoted as a dweller of an uncivilized forest in Russia who roared like a lion, hunted like a wolf and lived off berries, Lionel was deemed by the show's presenters the "missing link." In reality, Lionel had a medical condition called hypertrichosis, which produced excessive facial and body hair. He was raised as a performer in European sideshows with his father, who had the same condition. When Lionel reached adolescence, he became an avid reader. As a result, his public persona transformed from a savage child, or a "missing link," to an upper-class, educated, aristocratic young man. This was clearly at odds with eugenic principles, which asserted that the growing number of people like Lionel could outbreed the eugenically fit population, leading to chaos and the downfall of American society.[183]

Charles Davenport and his fellow eugenicists expressed grave concerns about genetic mixing between races, which they believed might produce primitive, animal-like qualities. As such, they were particularly interested in studying Coney Island performers Toney, the "Alligator Skin Boy," and Susi the "Elephant Skin Girl." These two suffered from ichthyosis, a genetic disorder that covers the body with tough, dark, rectangular scales.

Another performer studied was Prince Randian, also known as the Living Torso, a native of New Guinea who was born with phocomelia, a congenital condition that involves malformations where the human arms and legs result in flipper-like appendages. In the climax of the 1932 film *Freaks*, Randian is featured sinisterly crawling through the mud with a knife in his mouth. Despite his portrayal in the film and at Coney Island,

Randian was married and raising five perfectly normal children, a fact that directly undermined the misguided notions of eugenics.

During the nineteenth century, there was a strong fascination with jungle-like primitivism and stage personifications of dark savages, who were portrayed using exotic costumes and other imagery.[184] Eugenicists were naturally drawn to this phenomenon and were keenly interested in studying the cranium size of these individuals to assess their intelligence. Davenport noted, "There are structural differences that are innate and racial and need to be measured and studied."[185] Among these subjects, the ERO was particularly fascinated with "Chief Pantagal," who performed at Coney Island with live snakes and chickens before biting off their heads, chewing and swallowing them as part of a racialized performance of a savage jungle man.[186] Hoping to connect the eugenic dots, ERO agents collected a sample of Chief Pantagal's hair and sketched a brief pedigree chart that included his wife, sisters and daughter, all of whom had the same hair type.

In an effort to portray the differences between high and low cultures, ERO agents also studied "Chief Amok," a "Bantos head hunter" who often dressed in tribal regalia. Another performer, known as "Little Egypt," was a scantily clad African woman who danced provocatively for crowds of men. Eugenicists associated race and sexuality through the portrayals of Chiefs Pantagal and Amok, along with others at the Coney Island exhibits. The ERO's studies were used to reinforce Davenport's belief that Africans have "a strong sex instinct without corresponding self-control."[187] The racist notions derived from these performances inspired eugenicists to search for the scientific "truth" through displays of Black bodies in primitive and sexualized situations.[188] ERO officials also used these studies to further promote the authority, legitimacy and need for eugenics.

Many of the exhibits at Coney Island were portrayed as inverted beauty pageants, which eugenicists also sought to exploit. Mary Ann Bevan was billed as the "Ugliest Woman in the World." She suffered from acromegaly, a pituitary tumor that caused abnormal bone growth in her face and limbs. This left her with a masculine and distorted appearance. Grace Gilbert, known as the Bearded Lady, along with "Fat Ladies" and "Midget Madames," was similarly presented.

Spectators were also entertained by "Lilliputia," a municipality with approximately three hundred "midgets." ERO agents often photographed these performers alongside so-called normal individuals to accentuate the eugenics-based propaganda that society would be in danger if these

Chief Pantagal, a circus sideshow performer at Coney Island, New York. *Photo courtesy of the American Philosophical Society.*

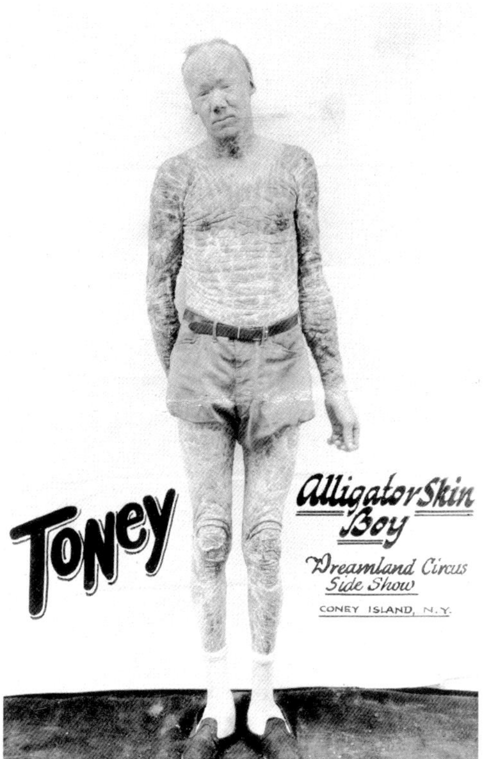

Left: Toney, the "Alligator Skin Boy," performer at the Dreamland Circus Side Show, Coney Island, New York. *Photo courtesy of the American Philosophical Society.*

Below: Lionel, the "Half Man/Half Lion," performer at the Dreamland Circus Side Show, Coney Island, New York, December 18, 1928. *Photo courtesy of the American Philosophical Society.*

"freaks" continued to spread their defective genes. Despite these efforts, eugenicists could never explain how most of these individuals, particularly the women, had little trouble finding husbands, raising children and living seemingly normal lives.

Davenport and his ilk had once again found an isolated community to be used for eugenic purposes. By infiltrating and exploiting the wildly popular Coney Island freak shows, they attempted to portray a world of genetic chaos unless the disciplined science of eugenics was adopted. Such exploitation continued until 1935, when the last photographs were taken.

Prior to the official opening of the Eugenics Record Office in 1910, eugenicists were well on their way toward increasing the rate of institutionalization of those eugenicists declared "defectives." The creation of the new category of "feeblemindedness" and other eugenics-based diagnoses helped accelerate this goal in a dramatic fashion.

Harry Laughlin studied U.S. census records to determine the number of inmates in various institutions including prisons, psychiatric facilities and poorhouses, as well as the infirm.[189] According to his calculations—based on the eleventh U.S. census, in 1890—the total U.S. population was 62,222,250, of which 369,064 were inmates in various institutions with 5,254 of them being declared "feebleminded." However, when the thirteenth census was conducted just twenty years later, it was revealed that the overall U.S. population had grown to nearly 92 million. The number of inmates at various institutions had ballooned to 841,244, and the number of those diagnosed as feebleminded—20,731—had nearly quadrupled. Moreover, in the same amount of time, the number of facilities that focused on treatment for feeblemindedness nearly tripled.

There may be reason to question the veracity of Harry Laughlin's data. As a staunch eugenicist and second-in-command at the ERO, he had every incentive to embellish his calculations to fit the eugenic narrative. However, eugenicists enjoyed unfettered access to most of the state-run institutions, which freely provided records of their populations. According to Laughlin's data, there is no doubt that the remedies initially proposed in 1914 by the eugenic section of the American Breeders Association, particularly for people with the manufactured condition of feeblemindedness and other similar diagnoses, was well entrenched and yielding the results that eugenicists had hoped for.

Officials at the Eugenics Record Office went to great lengths to target, and foster the mass institutionalization of, so-called defectives during their peak reproductive years. However, this was only the beginning. If they hoped to reach their goal of eliminating the perceived genetic defects of millions more Americans from the population, far more draconian measures, both legal and medical, would be required—and they were soon to be implemented on a massive scale.

Chapter 8

THE QUEST FOR MARRIAGE RESTRICTION

Idiots, imbeciles, and degenerate criminals are prolific, and their defects are transmissible. Each person is a unit of the nation, and the nation is strong and pure and sane, or weak and corrupt and insane.
—F.W. Hatch, California Superintendent of State Hospitals, 1910

Throughout the first half of the twentieth century, a multitude of people were diagnosed with eugenic-based conditions like feeblemindedness and condemned to psychiatric facilities, largely during their peak reproductive years. However, this alone was not a strong enough measure for American eugenicists, who had long sought to eliminate the so-called defectives of society. One additional measure that they proposed included the implementation of eugenic-based marriage restriction laws in each state.

On March 2, 1910, after soliciting commentary from medical professionals in the state, the attorney general of California issued a lengthy legal opinion on the need for marriage restriction laws. F.W. Hatch, the superintendent of state hospitals in Sacramento, stated:

> *Idiots, imbeciles, and degenerate criminals are prolific, and their defects are transmissible. Each person is a unit of the nation, and the nation is strong and pure and sane, or weak and corrupt and insane, in the proportion that the mentally and physically healthy exceed the diseased. This grave danger has consumed the thought of great and good men in recent years. Much restrictive legislation has been suggested, and many states have*

> *passed marriage laws for the purpose of regulating, as far as possible, the propagation of degenerates through the marriage relation.*[190]

Despite the widespread optimism of eugenicists, it was clearly understood that this measure would only yield limited results. Hatch went on to state, "Unfortunately, matrimony is not always necessary to propagation, and the tendency of these several different laws is to restrict procreation only among the more moral and intelligent class, while the most undesirable class goes on reproducing its kind."[191]

Officials at the ERO, including Harry Laughlin, were particularly skeptical of the overall effectiveness of marriage restriction laws. He noted,

> *Restrictive marriage laws and customs will have but little effect upon the socially inadequate classes. This is amply demonstrated by Davenport in Bulletin Number Nine of the Eugenics Record Office: "State Laws Limiting Marriage Selection Examined in the Light of Eugenics." For persons of sound mind and morals, but suffering from severe hereditary handicap, these remedies will be efficacious; but individuals are given the designation "socially inadequate" because, among other reasons, they are not amenable to law and custom.*[192]

Despite their limited expectations, eugenicists across the nation saw the marriage restriction laws as a necessary component of their mission. In New York, one of the earliest efforts to enact such a law occurred in February 1914, with a bill that was advanced in the New York State Assembly by Assemblyman Robert Lee Tudor. The proposed law included a eugenic marriage component, which required applicants seeking a marriage license in the state to first attain a physician's certificate stating that they were free from physical or mental diseases that were "likely to be contagious or hereditary."[193] Many medical professionals in New York fully supported eugenics-based marriage restriction laws in the state.

During the twentieth century, there was a great focus placed on the study and prevention of human blindness. This movement received a great deal of support and funding. Lucien Howe was widely known for his pioneering work on better vision for Americans. In 1876, he founded the Buffalo Eye and Ear Infirmary. He also helped thousands of newborn patients by bathing their eyes with drops of silver nitrate to fight neonatal infection, a practice that became law in New York in 1890. In 1918, Howe, who was widely known for his work in the United States and Europe, was elected to serve as president of the American Ophthalmologic Society. He also helped fund the Howe Laboratory of Ophthalmology at Harvard University. Howe was also a staunch eugenicist who ultimately became president of the Eugenics Research Association.[194]

Despite widespread sympathy for those with the condition, eugenicists believed that blindness was a hereditary defect that could be dealt with only through marriage restriction, segregation and sterilization.[195] Thus, they zealously sought to implement eugenic measures affecting those afflicted with blindness. At the onset, they began to compile a list of all known blind people in the nation and sought to introduce sweeping eugenics-based marriage restriction laws, along with proposing other options, such as creating segregated colonies of the blind or sterilization.

On April 5, 1921, Lucien Howe introduced bill no. 1597 to the New York State senate to amend the state's Domestic Relations Law. The proposed law sought to require "the town clerk upon application for a marriage license to ascertain as to any defects in either of such applicants, or in a blood relative of either party." If such defects were found, the town clerk or any taxpayer was empowered to appeal to a local judge, who would have the authority to appoint an ophthalmologist and a eugenic doctor, whose testimony could be used to prohibit the marriage.[196] That same year, Howe also proposed New York State Assembly Bill No. 605, which sought to establish marriage bonds that would require couples to pay as much as $14,000 ($130,000 in today's money), which would be forfeit if they were to be declared "unfit." Interstate deportation was also proposed in the law—that is, to send all those declared to be defective back to their home states.

All the proposed New York laws failed, but Howe's eugenic efforts regarding blindness never ceased. He partnered with attorneys associated with Columbia University to draft legislative language that could withstand legal scrutiny. He frequently corresponded with Charles Davenport and Harry Laughlin at the ERO to further their plans to compile a list of all the blind patients in New York State. They even discussed means to pressure the New York State Board of Health to dispatch field-workers to track down families with blindness. Despite having numerous connections with eugenicists, including New York State Commissioner of Public Health Hermann M. Biggs, an official program to hunt down all of those with blindness in the state was never established.[197]

These legislative setbacks did not deter Howe, who remained committed to having a marriage restriction law for the blind passed in New York. On February 1, 1926, he sponsored a modified version of his original bill in the

New York State Assembly, which included the requirement for all applicants to swear to the following statement: "Neither myself nor, to the best of my knowledge and belief, any of my blood relatives within the second degree have been affected with blindness."[198] The bill also proposed significant cash bonds that would make it almost financially impossible for blind people to marry in the state. Ultimately, the bill was never enacted.

Lucien Howe was praised for his tireless efforts to eliminate blindness, which continued until his death on December 17, 1928, at the age of eighty. He was fondly eulogized by his fellow eugenicists as a "true gentlemen, a broad scholar…[who] loved his fellow men." For eight decades, the American Ophthalmological Institute has awarded the Lucien Howe Medal for service to the profession and mankind.[199] Despite the legislative setbacks in New York, eugenicists across the country continued to grapple with ways to enact meaningful marriage restriction laws. They would eventually find success in the state of Virginia.

Walter Ashby Plecker (1861–1947) was a physician and official at the Registrar of Vital Statistics in Virginia. He received a medical degree at the University of Maryland at Baltimore and studied obstetrics at the New York Polyclinic. He later opened a practice in Virginia and began working with family records before settling in Elizabeth City County, Virginia, a historic county known for maintaining meticulous records dating back to the 1630s. In 1900, the county established a health department, which maintained birth and death records. A few years later, Plecker was hired as the county health officer, and in that capacity, he carefully recorded local births and deaths.

Plecker was a racist who was consumed with the purity of the White race. As a result, he would become one of the United States' leading recorders of demographic information on Blacks, Native Americans and other people of color.[200] He was also a staunch eugenicist and deeply troubled by the mixing of blood between Whites and people of color. In an official pamphlet he created for the Virginia State Health Bureau, he stated:

> *The white race in this land, is the foundation upon which rests its civilization, and is responsible for the leading position which we occupy amongst the nations of the world. Is it not, therefore, just and right that this race decide for itself what its composition shall be, and attempt, as Virginia has, to maintain its purity?*[201]

Plecker particularly lamented over the amount of racially mixed marriages in Virginia. He became obsessed with the increasing "Negro" population and was determined to take measures to curb it. By the age of fifty-one, Plecker had been hired as the county registrar and had unfettered access to records dating as far back as three hundred years.

Despite lacking scientific expertise, Plecker was eager to assist in the eugenics cause. He was a proud member of the American Eugenics Society and a staunch supporter of scientific racism. Plecker also regularly corresponded with Charles Davenport and Harry Laughlin and shared with them his belief that in Virginia, one's ancestry was either completely White or it wasn't—in other words, if a person possessed mixed blood, then they were not White.

In 1921, Plecker implemented strict rules for caretakers, midwives, obstetricians, town clerks and clerics of Virginia, obligating them to properly report all birth, marriage or death records of those they interacted with. Over the next few years, he meticulously catalogued the data of more than a million births and deaths since 1912, along with thousands of marriages annually. He also coordinated with the ERO to create the United States' first statewide eugenic registry.[202] His ultimate goal was to preserve the purity of the White race, and the best method he could fathom was statewide legislation forbidding interracial marriage, which both he and officials at the ERO had called the "mongrelization" of Virginia's White race.

To halt this so-called mongrelization, powerful Whites in Virginia organized a campaign to ban marriages in the state between a certified White person and anyone with even "one drop" of non-Caucasian blood.[203] This quest drew high praise from many of the leading "raceologists" of the time. Margaret Sanger, the controversial birth control activist, noted, "I consider such legislation…to be of the highest value and greatest necessity in order that the purity of the white race be safeguarded from possibility of contamination with nonwhite blood.…This is a matter of both national and racial life and death."[204] On March 8, 1924, the state legislature in Richmond passed the Racial Integrity Act of 1924.

Once the law was enacted, Plecker immediately distributed health bulletins with strict instructions attached to all marriage applications in the state. One such bulletin stated, "As color is the most important feature of this form of registration, the local registrar must be sure that there is no trace of colored blood in anyone offering to register as a white person. The penalty for willfully making a false claim as to color is one year in the penitentiary."[205] Plecker also sent threatening correspondence from the Bureau of Vital Statistics to new parents, newlyweds, midwives and physicians warning them of potential criminal exposure if they violated the law. In a letter dated

April 30, 1924, Plecker wrote to Mrs. Robert H. Cheatham of Lynchburg, Virginia, stating:

> *We have a report of the birth of your child, July 30th, 1923, signed by Mary Gildon, midwife. She says that you are white and that the father of the child is white. We have a correction to this certificate sent to us from the City Health Department at Lynchburg, in which they say that the father of this child is a negro. This is to give you warning that this is a mulatto child and you cannot pass it off as white. A new law passed by the last legislature says that if a child has one drop of negro blood in it, it cannot be counted as white. You will have to do something about this matter and see that this child is not allowed to mix with white children. It cannot go to white schools and can never marry a white person in Virginia. It is an awful thing.*[206]

Over time, Plecker worked tirelessly to enforce the law, which included the reporting of any suspected violations to law enforcement, decertification of mixed-race marriages and prevention of school admittance for mixed-race children. He also pressured operators of White cemeteries to report the deaths of those suspected of having "Negro" bloodlines. He opined that, "To the white owner of a lot, it might prove embarrassing to meet with negroes visiting at one of their graves on the adjoining lot."[207]

Eugenicists across the country hailed Walter Plecker as a hero for helping pass the Racial Integrity Act in Virginia. The rights of millions of Americans to freely marry in the state were directly affected by his actions, which were driven by virulent racism and White supremacy, all shrouded under the guise of eugenics. After establishing the framework for eugenics-based marriage restriction and eventual immigration laws, eugenicists set their eyes on the far more sinister, irreversible and widespread measure of mass sterilization.

Chapter 9

THE PURSUIT OF MASS STERILIZATION

The great mass of humanity is not only a social menace to the present generation, but it harbors the potential parenthood of the social misfits of our future generations.
—Harry Laughlin, 1914

Long before the opening of the Eugenics Record Office in 1910, Charles Davenport was busy establishing and promoting the new pseudoscience of eugenics. As an active member of the American Breeders Association (ABA) Committee on Eugenics, he urged members of the public to submit their family histories so that he could search for unusual genetic conditions and defects.[208] This information was used to create a multitude of pedigree charts, which contained data on the lineages of entire families. At a subsequent committee meeting, Davenport informed his colleagues that six hundred pedigree charts had been created, which provided ample evidence that feebleminded "defectives should be restrained from passing on their condition."[209]

Harry Laughlin was in agreement with and fully committed to Davenport's work. He stated, "It would be possible theoretically to sterilize wholesale those individuals thought to carry defective hereditary traits, and thus at one fell stroke cut off practically all of the cacogenic varieties of the race."[210] Laughlin adopted this sentiment as a mandate and proposed the appointment of a committee to study the "experiment in sterilization" made by the few states that had enacted legislation in this field. At the direction of ABA president Bleecker Van Wagenen, the group began to compile data on sterilization in 1911, and three years later, Laughlin presented the infamous *Report of the Committee of the Eugenic Section of the American Breeders' Association to Study and to Report on the Best Practical Means for Cutting Off the Defective Germ-Plasm in the Human Population*. Laughlin would ultimately use this data to develop a model eugenic sterilization law for other states to use in order to enact similar legislation that could withstand constitutional scrutiny.

This data compiled by Laughlin was first presented at the first International Eugenics Congress in London in 1912.[211] Those in attendance included well-respected physicians and genetic consultants such as Raymond Pearl and Lewellys Barker of Johns Hopkins University, Nobel Laureate and surgeon Alexis Carrel of the Rockefeller Institute, Yale economist Irving Fisher and Henry Goddard of the New Jersey Vineland Training School for Feeble-Minded Girls and Boys. At the conference, Van Wagenen declared that people of "defective inheritance" should be "eliminated from the human stock."[212] The group calculated that approximately 10 percent of the U.S. population was "totally unfitted to become parents of useful citizens." Charles Davenport was also in attendance and stated, "The only way to prevent the reproduction of the feeble-minded is to sterilize or segregate them. Law must take lessons from biology."[213]

As the ERO continued to create pedigree charts from the massive amounts of family data it had collected, it publicly declared, in 1913, a set of eugenic goals for the organization, which included "the study of America's best blood-lines…and the study [of] the best methods of restricting the strains that produce the defective and delinquent classes of the community."[214] One year later, Harry Laughlin spoke at the First National Conference on Race Betterment, hosted by Dr. J.H. Kellogg in Battle Creek, Michigan. The audience was eager to hear a summary from the ERO of its plans to eliminate "the great mass of defectiveness…menacing our national efficiency and happiness."[215] Laughlin surmised that the only way to cleanse the human "breeding stock" would be to sterilize fifteen million Americans over a period of sixty-five years. He added that the mass sterilization program would begin in institutions for approximately ten years before expanding to the public at large.

In February 1914, the Eugenics Record Office published Bulletin No. 10, which contained information on Harry Laughlin's voluminous report, complete with data and commentary in support of eugenic sterilization and other harsh eugenic measures. In summarizing the scope of the committee's work, Laughlin stated, "The great mass of humanity is not only a social menace to the present generation, but it harbors the potential parenthood of the social misfits of our future generations." He argued that the "socially inadequate" groups must be cut off and that "this is the natural outcome of an awakened social conscience; it is in keeping not only with humanitarianism, but with law and order, and national efficiency."

Eugenicists understood that plans for the mass sterilization of millions of Americans would be met with sharp criticism and public fear. They believed that the only way to ensure the durability of the plan was to establish laws in each state that could sufficiently withstand constitutional scrutiny. As a result, Harry Laughlin began to create a model sterilization law that each state could replicate in its own legislature. The preamble outlined the law's objective as follows:

> *An Act to prevent the procreation of feebleminded, insane, epileptic, inebriate, criminalistic and other degenerate persons by authorizing and providing by due process of law for the sterilization of persons with inferior hereditary potentialities, maintained wholly or in part by public expense.*[216]

The law also proposed the establishment of a eugenics commission in every U.S. state to examine those who were institutionalized and the public at large.

In 1922, Laughlin released a book titled *Eugenical Sterilization in the United States*, which compiled information on all the eugenic sterilization laws that had been passed in each state, along with legislative and judicial commentary on each law. The book was hailed by eugenicists around the world, and in the United States, it presented a blueprint for the implementation of mass eugenic sterilization across the country. The ERO was fully committed to realizing this goal.

Prior to the opening of the ERO, several states had already passed compulsory sterilization laws.[217] The first state, both in the nation and the world, to pass a eugenic sterilization law was Indiana, in 1907. This law targeted "confirmed criminals, idiots, imbeciles, and rapists."[218] The State of Washington also passed a sterilization law in 1909, followed by a similar law in New Jersey in 1911, which targeted the feebleminded, epileptics, those convicted of rape and habitual criminals.[219] Iowa and Michigan passed similar laws in 1914 and 1916, respectively, along with several other states.

However, many of these laws were ultimately struck down in appeals courts for being unconstitutional.[220]

On April 16, 1912, New York enacted Chapter 445 of the Public Health Law.[221] This eugenic sterilization law was designed "in relation to operations for the prevention of procreation." The law targeted inmates of state institutions that housed insane, feebleminded and dependent inmates, along with confirmed criminals.

The new law methodically set forth the powers, duties and compensation of a board of examiners to exercise their duties, which included an examination of the mental and physical records of patients held at state institutions. The heart of the law was found within Section 351, which gave the board the authority to determine if sterilization was appropriate. The relevant portion of that section stated:

> *If, in the judgment of a majority of said board procreation by any such person would produce children with an inherited tendency to crime, insanity, feeble-mindedness, idiocy, or imbecility, and there is no probability that the condition of any such person so examined will improve to such an extent as to render procreation by any such person advisable, or if the physical or mental condition of any such person will be substantially improved thereby, then said Board shall appoint one of its members to perform such operation for the prevention of procreation as shall be decided by said board to be most effective.*[222]

Inmates within these institutions who had been convicted of rape or multiple other offenses were automatically deemed to possess the criminal tendencies covered by the law and thus could be sterilized against their will. Moreover, while the law forbade any illegal or non–board authorized operations, the penalty for such offenses was merely a misdemeanor, and thus of limited deterrence. Lastly, while the statute provided for the appointment of counsel for those to be subjected to sterilization, it is unclear how any viable defense could have been made to prevent compulsory sterilization in the face of such clear and sweeping statutory language.

Numerous officials at state institutions throughout New York were staunch supporters of eugenics and fully embraced the sterilization law. In February 1918, Dr. C.A. Potter, the superintendent of Gowanda State Hospital in Collins, New York, which accounted for more than half of the sterilization procedures in the state, noted:

> *The public should be shown that insane, epileptics, feeble-minded and criminals have no right to procreate, from an economic standpoint as well as from the point of eugenics. The insane, epileptics, feeble-minded and criminals of child-bearing age should be sterilized.*[223]

Potter believed that the law would be "of great value in preventing recurrence of attacks in insanity" and "its eugenical value would be greater than any law of recent years which applies to institutions."

Overall support from medical professionals and other state officials in New York was widespread. O.M. Grover, a resident physician at the Reformatory for Women in Bedford Hills, New York, commented, "I think all mental defectives who are custodial charges should be sterilized." Katharine Bement Davis, who directed the New York Parole Commission, Dr. Edward Humphries at the Letchworth Village Psychiatric Hospital and Henry Goddard of the Vineland Training School for Feeble-Minded Boys and Girls all approved of the law.

In February 1918, Dr. John Ross, medical superintendent of Dannemora State Prison in Clinton, New York, supported the sterilization procedure and stated, "There is no doubt in my mind that this operation, if carried out extensively among the insane, feeble-minded and certain of the criminal type, would be of great eugenical value." Dr. Frank Heacox, a physician at the Auburn State Prison in upstate New York, opined that the statute was eugenically invaluable. Dr. Arthur Ward, superintendent of the State Hospital in Buffalo, added that the law "may be of a great deal of value in selected cases, as childbearing, for instance, brings on recurring acts of insanity. Eugenically, the statute is of much value in preventing the propagation of defectives."[224]

On Long Island, Dr. William A. Macy, superintendent of Kings Park State Hospital, stated in January 1918 that with sufficient public support behind the law, "it should prove of value, especially in paroled or discharged cases of chronic insanity, mental deficiency and frequently occurring cases of mental disorder."[225] In March 1918, Dr. G.A. Smith, superintendent of the Central Islip State Hospital, opined that the law had eugenical value in cases of idiots, imbeciles, mental defectives and epileptics confined in institutions.

With carefully crafted language, the State of New York legalized compulsory sterilization based purely on eugenic criteria. It became just the third state at the time to have such a law on the books, joining Washington and New Jersey. Despite having a eugenic statute codified into New York law that, under certain conditions, permitted the sterilization of patients held in state institutions, eugenicists understood that the law was always subject to litigation and possible repeal. What they desired was a test case to determine if the law could withstand legal scrutiny. Just two years after the law was enacted, the first test case presented itself in New York when the Rome Custodial Asylum in upstate New York sought to sterilize a twenty-two-year-old man named Frank Osborn. This was one of several cases brought by eugenicists nationwide with the ultimate goal of having one of them reach the United States Supreme Court so that the highest court in the land could, once and for all, establish the constitutionality of eugenic sterilization.

On September 17, 1915, the trial of Frank Osborn began. Osborn was not a criminal, and this was not a civil trial. Instead, the case was brought against him by the board of examiners of the Rome Custodial Asylum to gain court approval to forcefully sterilize Osborn.[226] According to court filings, Frank Osborn was a twenty-two-year-old inmate of the Rome Custodial Asylum in Rome, New York, where he had been confined for eight years.[227] Although Osborn was physically strong, the results of a Binet intelligence test indicated that he was feebleminded, with the mental capacity of an eight-year-old child. In support of the board of examiners' case, Lemon Thomson, a doctor and one of the three members of the board, testified about the Osborn family lineage:

> *After a careful examination by the Board, we have learned that said Frank Osborn comes from a family of degenerates. He is one of sixteen children, eight of whom are dead. Five brothers and sisters besides himself are confined in State institutions for the feeble-minded; one, a feeble-minded brother, lives with a farmer and is intemperate, incapable, and untrustworthy; one sister, the brightest of the family, lives with and keeps house for a man to whom she is not married, though she has a husband living. She is immoral and has been an inmate for two years of a house of prostitution. Of his dead brothers and sisters, one died in an institution for feeble-minded and seven died before becoming one year of age. The father of said Frank Osborn was feeble-minded and the son of a man who was an epileptic and who lost his mind before death. Said Frank Osborn's mother is living, is feeble-minded and comes from a family of defectives. Her mother was feeble-minded, and one sister and two brothers of Frank's mother were feeble-minded.*[228]

The board also admitted evidence showing that the Osborn family had caused economic strain on the county, costing the county what the board estimated to be approximately $10,000 since the family became charges of the state.[229]

The Osborn litigation did not produce the result that the eugenicists hoped for. On March 5, 1914, the Supreme Court of Albany County deemed that the law was unconstitutional. In the decision, Justice William P. Rudd reasoned that "the provisions of the Federal Constitution, to which this law is offensive, is that part of the Fourteenth Amendment which declares that no state shall deny to any person within its jurisdiction the equal protection of the laws." The court perpetually enjoined and restrained the defendants from performing or permitting to be performed the sterilization procedure in New York State.[230]

Under appeal, this decision was upheld by the Supreme Court of Albany County on the same constitutional grounds, setting the stage for a final appeal before the New Yorks Court of Appeals. However, the test case that eugenicists hoped would lead to the implementation of eugenic sterilization in New York was thwarted when the New York sterilization statute was unanimously repealed on May 10, 1920, just eight years after it was enacted, rendering the constitutional question of forced sterilization in New York moot; the appeal was withdrawn.[231]

From the law's enactment in 1912 until it was repealed in 1920, forty-one women and one man were sterilized in various New York institutions. Of these surgical procedures, one was a vasectomy performed at the Auburn State Prison in Auburn, New York.[232] Twelve salpingectomies were performed by the Buffalo State Hospital, and twenty-four salpingectomies and five ovariotomies were performed by the Gowanda State Hospital at Collins, New York.[233] However, of all forty-two sterilization procedures performed in New York while the statute was in force, none of them were compulsory. Instead, they all were performed by special arrangements made between doctors and the patients and their families under the laws and customs governing ordinary surgical operations.[234]

Despite garnering an abundance of support from medical practitioners in numerous state-run facilities, along with all the resources provided by the Eugenics Record Office, the eugenical sterilization law was short-lived and completely ineffective. The legislative setback in New York appeared to be a severe blow to the ERO, which was housed in the same state. Laughlin summarized his frustration with the legal process as follows: "In short, the history of this law in New York State is a record of politics, incompetency, and discredit. It has set back eugenical progress among the state's institutions more than ten years."[235] Despite the loss, it was only a matter of time before a eugenic sterilization law case would be brought before the highest court. Moreover, eugenicists remained optimistic over the increasing rates of sterilization in other states. According to Laughlin's data, as of January 1921, a total of 3,233 eugenic sterilizations had been performed in fifteen states that had sterilization laws. California led the way by far with 2,558 procedures, followed by Nebraska with 155, Oregon with 127, Indiana with 120 and Wisconsin with 76.[236]

Despite the data, eugenicists still required a test case that could be brought before the United States Supreme Court. This was the ultimate goal of eugenicists nationwide, who keenly understood that a ruling by the high court upholding a state eugenic sterilization law would pave the way for a nationwide legislative campaign in which all states could lawfully sterilize "defectives" in untold numbers. In less than a decade, one such case was formed when the State of Virginia sought to sterilize a young woman, who was declared feebleminded, against her will.

Chapter 10

BUCK V. BELL

Supreme Injustice

It is better for all the world, if instead of waiting to execute degenerate offspring for their crime, or to let them starve for their imbecility, society can prevent those who are manifestly unfit from continuing their kind.... Three generations of imbeciles are enough.
—Chief Justice Oliver Wendell Holmes Jr., United States Supreme Court in Buck v. Bell, *May 2, 1927*

One of the most enduring achievements ever reached in the name of eugenics came in the form of judicial approval of its policies. For the purposes of eugenic sterilization, there was no greater achievement or validation than the United States Supreme Court decision in 1927 that allowed the State of Virginia to sterilize a young woman against her will. In a decision that has never been reversed, the highest court in the U.S. judicial system provided full endorsement to eugenicists across the country to forcefully sterilize of thousands of Americans. It also provided the ultimate legitimacy to a field of science that, within a mere thirty years, would be thoroughly discredited. This chapter explores the infamous *Buck v. Bell* case.

Carrie Elizabeth Buck was born on July 2, 1906, in the rural county of Charlottesville, Virginia. Mired in poverty, her mother was forced to place Carrie in foster care when she was just a toddler. John Dobbs and his wife, Alice, agreed to take Carrie into their home in downtown Charlottesville. Although theirs was a loving home, Carrie was treated more as a housemaid than a daughter, and the couple removed her from school and forced her to perform many household chores. In 1924, the Dobbs family was infuriated to learn that the seventeen-year-old Carrie was pregnant out of wedlock, and they evicted her from their home.[237]

John Dobbs made an appointment with Mary Duke, secretary of public welfare. Although Duke never personally met Carrie, she knew Carrie's mother, Emma, whom she believed to be of "bad character." After further discussion, Mary Duke agreed to assist. However, instead of placing Carrie in a home for unwed mothers, she arranged for her to be institutionalized.[238] A short time later, Carrie gave birth to a daughter named Vivian.

Mary Duke arranged to have Carrie placed at the Virginia Colony for Epileptics and Feebleminded (hereinafter referred to as the colony). At the time, Dr. Albert Priddy was superintendent of the colony. He was also an ardent supporter of eugenics who once described epileptics as among "the most pitiful, helpless, and troublesome of human beings."[239] Priddy strongly favored institutionalization, sterilization and marriage restriction laws in the name of eugenics. Even before they were legalized in the state, Dr. Priddy appears to have performed eugenical sterilization procedures. He once performed an operation on a young woman "for relief of a chronic pelvic disorder, which sterilized her." The so-called pelvic disorder was diagnosed in dozens of women at the facility, and the procedure used to treat it also caused them to be sterilized.[240]

Willie Mallory was a mother of nine children who was placed under the control of the colony. In 1917, she sued Dr. Priddy for "wrongful and illegal assaults and batteries" after he anesthetized and sterilized her. At trial, Dr. Priddy claimed that Mallory gave him informed consent for the procedure and the sterilization was an unintended result. In March 1918, a jury ruled in his favor, but the judge in the case provided a stern warning to Dr. Priddy not to conduct any further procedures on inmates until a proper law was in place.[241] This warning against sterilization also applied to all other hospital administrators in the Virginia.

In 1924, the State of Virginia passed its own eugenic sterilization law. Dr. Priddy and other officials at the colony had long been advocating for such a law and enlisted powerful attorney and former state senator Aubrey Strode for help with its passage. In August 1924, Dr. Priddy provided a list of potential patients to be sterilized. A month later, the colony hired Strode as its counsel. As plans for a test case were being formulated, Carrie Buck had already been at the colony for three months. Since her mother, Emma, had been declared feebleminded and Carrie had recently given birth to a daughter, Carrie was seen as an ideal candidate for the test case.[242] A short time later, the board of the colony voted to sterilize Carrie. Its resolution stated,

> *Carrie Buck is a feebleminded inmate of this institution and by the laws of heredity is the probable potential parent of socially inadequate offspring, likewise afflicted, that she may be sexually sterilized without detriment to her general health, and that the welfare of the said Carrie Buck and of society will be promoted by such sterilization.*[243]

At the time, eugenicists nationwide were grappling with a rash of judicial setbacks. The eugenical sterilization laws in New York, New Jersey, Iowa, Michigan, Nevada, Indiana and Oregon were all struck down for being unconstitutional. Then, on September 30, 1924, Harry Laughlin received a letter from Aubrey Strode. The letter praised Laughlin for his work on eugenics and explained how his book *Eugenical Sterilization in the United States* was pivotal in the drafting of Virginia's sterilization law. Strode explained the case against Carrie Buck and asked if Laughlin would be willing to testify as an expert witness.[244]

Strode's letter contained significant factual inaccuracies. First, he described Carrie as a nineteen-year-old "feeble-minded mother," even though she had just turned eighteen and was never medically diagnosed as feebleminded. He also declared that Carrie's mother was feebleminded and that Carrie was "the mother of a feeble-minded child," Vivian. Given the dubious nature of the intelligence testing used at the time, it could not be certain that Carrie's mother, Emma, was indeed feebleminded, and Carrie's daughter, Vivian, who was just six months old at the time, was never tested in any way.[245] Even though the facts relayed to him were largely false, Laughlin was eager to assist. He explained, "The key was that both the immediate ancestor and the immediate offspring were feebleminded."

Laughlin agreed to assist and mailed copies of a "family tree folder" to Strode's team along with a set of "single traits sheets," which were both useful tools devised by the ERO to search for genetic defects. However, tracing Carrie's lineage proved to be difficult. Priddy explained to Strode that "this girl comes from a shiftless, ignorant and moving class of people," thus rendering it impossible to get accurate data on the family.[246] Laughlin never received the completed forms on the Buck family. Nor did he ever physically meet with Carrie, her mother or her daughter. Nevertheless, he pledged his support, relying solely on the false or incomplete information provided to him.

With Laughlin unable to travel to Virginia to testify in person, Strode sent him written interrogatory questions to be answered and used at the trial. Laughlin promptly completed the questions and had the document notarized in Cold Spring Harbor before mailing it back to Strode. Laughlin described Carrie as having "mental defectiveness, evidenced by failure of mental development." Although she was eighteen years old, she possessed the mental capacity of a nine-year-old according to the Stanford revision of the Binet-Simon test. Laughlin added that Carrie maintained a record of "immorality, prostitution and untruthfulness." Lastly, he described Carrie as having a "rather badly formed face."[247]

Without ever meeting any of the family members, Laughlin described Carrie's mother, Emma, who was fifty-two years old at the time, as a woman with the mental capacity of a child less than twelve years old. As he did with Carrie, he described Emma as having a record of "immorality, prostitution and untruthfulness." She had syphilis, he said, and had likely borne several illegitimate children, including Carrie.[248] He summarized the Buck family as follows: "These people belong to the shiftless, ignorant and worthless class of anti-social whites of the South."[249] Laughlin then declared Carrie's daughter Vivian feebleminded even though she was never tested in any way.

In a section titled "Analysis of Facts," Laughlin determined that Carrie's feeblemindedness was indeed hereditary. He explained that even though Carrie was moved from "the bad environment furnished by her mother" to a better environment provided by the Dobbs family, she showed no signs of improvement at school. This, he opined, was "typical…of a low-grade moron."[250]

A critical question for the trial was whether Carrie was likely to give birth to other defective children. On this point, Laughlin opined that given her own "feeblemindedness and moral delinquency," along with the evidence of hereditary defects in her family, Carrie should be considered a "potential parent of socially inadequate or defective offspring." Responding to the question of whether feeblemindedness could be transmitted to children, Laughlin cited his own book *Eugenical Sterilization in the United States* and asserted that the evidence showed "beyond a reasonable doubt" that it was indeed hereditary.

When asked to opine on the balance of the patient's interests against those of society, Laughlin answered that "modern eugenical sterilization is a force for the mitigation of race degeneracy which, if properly used, is safe and effective." Laughlin concluded his written testimony with a statement on the nature of the Virginia sterilization law, which he described as one of the most effective enacted statutes in the nation as it "avoided the principal eugenical and legal defects of previous statutes and has incorporated into it the most effective eugenical features and the soundest legal principles of previous laws."[251] Laughlin, unable to attend the trial, arranged for ERO field officer Arthur Estabrook to testify as an expert witness. The confidence of Strode and Dr. Priddy—now armed with the sworn written testimony of Harry Laughlin and a testifying witness from the ERO—was galvanized.

Aubrey Strode was one of Virginia's most prominent lawyers and a former state senator.[252] By the eve of the trial, he had already been practicing law for a quarter of a century. His experience, coupled with the importance of this case, compelled him to be fully prepared for trial. Strode had already procured a sworn statement by Harry Laughlin, and the Eugenics Record Office dispatched Estabrook to testify in person. This powerful combination would go a long way toward attaining the judicial authority to sterilize Carrie Buck.

In evaluating the case, Strode was concerned about the lack of testimony on the mental capacity of Carrie's daughter, Vivian. The importance of this testimony could not be understated. If Carrie, her mother, Emma, and her daughter, Vivian, were all declared feebleminded, then this would be definitive evidence that feeblemindedness was passed along through the three generations. As a result, it was decided that Estabrook would testify to Vivian's mental capacity.

As the trial date neared, anticipation was high. Eugenicists had long sought to present a case that would legalize eugenic sterilization and could ultimately withstand appeal before the U.S. Supreme Court, and they believed that this was such a case. Since appellate courts review only questions of law, it was critical for Strode to establish a damning and permanent evidentiary record against Carrie, leaving only questions of law to be decided on appeal. The only true way to accomplish this was to ensure that critical evidence would be admitted into the record without any significant objection from the opposing counsel. Regrettably, this was accomplished when Irving Whitehead was appointed to serve as Carrie's lawyer. The result was a complete miscarriage of justice.

Like Strode, Irving Whitehead was a prominent attorney in Virginia. He also served on the colony's board for fourteen years. Whitehead was so esteemed that a few months before the trial, a building was constructed at the colony in his honor.[253] During his tenure with the colony, Whitehead supported Dr. Priddy's persistent lobbying to legalize eugenic sterilization in the state, even describing the procedure as "therapeutic."[254] Moreover, just one week before the Buck trial gained the attention of the public, Aubrey Strode wrote a letter on behalf of Irving Whitehead to the Federal Land Bank of Baltimore recommending him for a general counsel position. These unmistakable connections would not be improper if Irving Whitehead was serving as an attorney for the colony in its quest to sterilize Carrie Buck. Amazingly, this was not the case. Instead, Whitehead, a strong ally of the colony and supporter of eugenical sterilization overall, was appointed to serve as Carrie's lawyer at the trial.

All attorneys are bound by a strict code of legal ethics to serve their client to the best of their ability, and any conflicts between the attorney's interests and those of their client must be openly declared. In this case, Irving Whitehead had a staggering number of legal conflicts with Carrie's interests that he never disclosed to her, which presented a miscarriage of his ethical obligations as an attorney. If Carrie, or anyone before the court, had knowledge of these conflicts, she or someone acting in her stead could have made objections before the trial began or perhaps have Whitehead removed as her counsel. Even Aubrey Strode, who was clearly aware of Whitehead's conflicts of interest, had a responsibility as opposing counsel to at least inform the presiding judge, but he neglected to do so. The stakes were simply too high and far more important than the rights of this poor young woman. Despite the obvious ethical violations, Whitehead's performance in the trial would ultimately determine whether he would provide adequate legal representation. Unfortunately, in both action and inaction, Whitehead's involvement in the case directly doomed Carrie Buck, as the record is rife with not only examples of his woefully inadequate representation but also statements he made that actually helped the colony's case against his client.

ON NOVEMBER 18, 1924, the trial of *Buck v. Priddy* began at the Amherst County Courthouse with Judge Bennett Gordon presiding.[255] The first witness presented by Strode was Anne Harris, a nurse from Charlottesville. She testified that she had known Emma Buck for more than a decade and described her as an "absolutely irresponsible" parent and a person "on the charity list for years" who was "living in the worst neighborhoods." She also opined that Carrie had a mental age of seven or eight.[256]

Any competent attorney uses cross-examination to discredit a witness testifying against his or her client. However, when Irving Whitehead cross-examined Harris, he neglected to ask her about the fact that she lacked the professional credentials to determine that Carrie had a low mental capacity. He also failed to elicit testimony that Carrie had successfully reached the sixth grade in school or that any deficiencies she possessed could have stemmed from poverty or lack of support. Instead, his line of questioning bolstered the colony's case against Carrie. For instance, during her questioning by Strode, Harris never mentioned that Carrie was exhibiting bad behavior. But during cross-examination, Whitehead elicited testimony from Harris that Carrie misbehaved in school by passing along notes to other children. He then asked if such conduct could constitute antisocial behavior, and Harris replied, "I should say so."[257] With Whitehead's direct assistance, the uncontroverted testimony that Carrie exhibited antisocial behavior was entered into the permanent record.

Strode then called former teachers of Carrie's relatives, who provided damaging testimony about their deficiencies, all of which went unchallenged by Whitehead. Strode also called Caroline Wilhelm, a social worker for the Red Cross who had accompanied Carrie when she was admitted to the colony. A month prior to the trial, Wilhelm told Strode that she had not reached any conclusion about the mental capacity of Carrie's daughter, Vivian. However, during the trial, Wilhelm testified that she'd evaluated Vivian just two weeks before and found her to be "not quite a normal baby." On cross-examination of Wilhelm, Whitehead failed to ask how she was—or if she even was—qualified to reach this conclusion. Even worse, he once again elicited information that was very damaging to Carrie's case. When being questioned by Strode, Wilhelm never stated that Carrie was immoral. However, on cross-examination, Whitehead specifically asked her: "Now, this girl, according to your viewpoint she has an immoral tendency?" Wilhelm replied "Certainly."[258]

Strode also called Mary Duke, the superintendent of public welfare at the time the Dobbs family sought to have Carrie committed, to testify. Although she'd had limited interaction with Carrie, Duke testified that "she didn't seem to be a bright girl." When questioned about Emma Buck, Duke testified that she "understood at the time she was of bad character."[259] Whitehead failed to make any attempts to undermine this testimony.

The next witness was Dr. Joseph S. DeJarnette, superintendent of the Western State Hospital in Virginia. He testified that in thirty-six years of practice, he had treated more than eleven thousand "mental defectives." He insisted that feeblemindedness was a "judicially ascertainable" condition and there were well recognized tests that would "safely classify those that are feebleminded." He added that feebleminded women like Carrie were "easily over-sexed" and that feeblemindedness was an inheritable trait. On this point, he opined, "I think Mendel's law covers it very well," even though he later stated that he did not possess a full understanding of Mendel's work. When asked for his opinion on Carrie, DeJarnette testified that he had heard she had an illegitimate child who "does not appear normal."[260]

DeJarnette offered glowing comments in support of Virginia's sterilization law and testified that the law promoted the "best interests" of the patient. He explained that if a patient in a state hospital were to be sterilized, they could then be "liberated" to the outside world "without bringing children into the world." He added that the law would benefit the interests of society because the "standard of general intelligence would be uplifted" and it would "lower the number of our criminals." He concluded his direct testimony by stating that Carrie's sterilization would benefit society overall.

There were numerous opportunities to undermine DeJarnette's testimony. For instance, since he admitted he did not fully understand Mendelian theory, it was not clear how he could competently have applied it to Carrie's case. Ultimately, Whitehead failed to raise DeJarnette's lack of credentials in hereditary science. Instead, he asked general questions about how prostitutes were "more or less feebleminded," an assertion DeJarnette agreed with. The remainder of DeJarnette's testimony went unchallenged, and once again, all this damning evidence became part of the trial record.

Strode then called Arthur Estabrook to the stand. Even though he was an employee of the Eugenics Record Office, he presented himself solely as a member of the scientific staff of the Carnegie Institute, and during his entire testimony, he refrained from making any reference to eugenics.[261] This appears to have been a tactical ploy by Strode's team to avoid any controversy being raised about the science of eugenics. Of course, as Carrie's attorney, Whitehead could have probed this further, but he neglected to do so in yet another example of how he was aligned with Strode in the case against Carrie.

Estabrook was first asked about Emma Buck, whom he confidently declared was feebleminded, along with many in her maternal line, which he described as the "Dudley germ plasm." When asked if he had the opportunity to evaluate Carrie to determine her mental capacity, Estabrook replied:

> *Yes, sir. I talked to Carrie sufficiently so that with the record of the mental examination—yes, I did. I gave a sufficient examination so that I consider her feebleminded.*[262]

Estabrook was also asked about the mental capacity of Carrie's daughter, Vivian. He stated, "I gave the child the regular mental test for a child of the age of six months, and judging by her reactions to the tests I gave her, I decided that she was below the average age for a child of eight months age." Estabrook never explained the nature of the test he administered to Vivian, and no follow-up questions were asked about it by Strode.

Perhaps more than with any other witness, there were many glaring opportunities to undermine Estabrook's testimony on cross-examination, and once again, Whitehead failed to pursue any of them. For instance, Estabrook's reply about whether he administered a test to Carrie was, at best, misleading. In reality, Carrie Buck was never administered any tests to determine her mental capacity, and Estabrook was the first person to ever claim to have tested her. Thus, at most, Estabrook's answer suggests that after reviewing the information presented to him secondhand, he surmised that Carrie was feebleminded. That is clearly not the same as him personally administering a test to her and testifying about the results of that test. Information about his failure to directly administer an intelligence test to Carrie could easily have been elicited from Estabrook, and if that admission was made, Whitehead could have pressed Estabrook on why he or anyone else failed to test her. This alone would have left a gaping hole in Strode's case, but once again, Whitehead neglected to even try. He also failed to challenge Estabrook's fieldwork methods and, most importantly, his conclusion that Vivian tested below average for a child of her age; he also did not ask about the type of test Estabrook administered to Vivian. All Estabrook's evidence went unchallenged and was admitted into the trial record, which could not be subjected to scrutiny later on appeal.

Strode's final witness was Dr. Priddy. After asking preliminary questions about his credentials, Strode asked the most critical question of the trial: why did the colony move to have Carrie sterilized? Priddy replied that Carrie had recently turned eighteen and that she would remain in the colony's custody for three decades, at a cost of approximately $200 per year for thirty years, which would deny her "all the blessings of outdoor life and liberty." He added that if Carrie was sterilized, she could find stable work and "probably find a man of her own level."[263]

Dr. Priddy was then asked to speak about Carrie's personal history. He replied that she was feebleminded, with a mental age of nine years old, making her a "middle-grade moron." This, of course, was a different diagnosis than "feebleminded," but Strode never sought to correct it, and Whitehead never bothered to challenge it. Priddy then testified that Carrie's mother, Emma, was feebleminded, with a mental age of about seven years and eleven months, based on the colony's testing, and added, "That meant two direct generations of feebleminded."[264] He stated that Carrie would be the mother of other defective offspring based on "the generally accepted theory of the laws of heredity." Priddy also spoke glowingly about the Virginia sterilization law, which he felt would "liberate" many of the patients at the colony. He described the law as "a blessing" for society. He then testified that many of the patients were eager to be sterilized because they "know it means the enjoyment of life and the peaceful pursuance of happiness…on the outside of institution walls."

Dr. Priddy added that he remained in contact with patients who had been sterilized and released from the colony. As began to share the story of two of these patients, he struggled to recall their names, saying, "Mr. Whitehead knows them both." Whitehead rose and, without being sworn in, stated, "Yes, put in there that I know them both." This exchange should have been astonishing to all present in the courtroom, as it clearly showed the alliance of Whitehead, Priddy and the colony. Nevertheless, no issue was raised by anyone, including Judge Gordon. As he did with the other witnesses, Whitehead failed to ask any meaningful questions or challenge any of the testimony presented by Dr. Priddy. Before closing the colony's case, Strode read Laughlin's sworn statement before the court, which was entered into the record in its entirety and without objection from Whitehead. The colony then rested its case, and the court adjourned for the day. It took just one day for Strode to present the colony's case and enter the necessary evidence into the record.

Now, the trial turned to Irving Whitehead to present a defense on behalf of his client. There were many opportunities for him to admit evidence that would undermine the case against Carrie Buck. Testimony from friends and family, including evidence that Carrie had already reached the sixth grade in school, would have helped rebuff the allegations made about her mental capacity. Carrie herself could have testified to dispel the notion that she was feebleminded and instead demonstrate the mental capacity of a capable young woman who wanted to have other children in the future. The defense also had the opportunity to solicit testimony from expert witnesses about the unreliability of the Binet-Simon test, as well as to undermine Estabrook's testimony and challenge the science of eugenics overall. Tragically, Whitehead failed to provide even a modicum of a case. Instead, he called no witnesses and refused to introduce a single piece of evidence on Carrie's behalf. After closing statements were made, the trial concluded.

In early February 1925, the Amherst County Circuit Court, unsurprisingly, announced a decision upholding the sterilization law and affirming the colony's request to sterilize Carrie Buck. Judge Gordon ruled that the sterilization law was constitutionally valid. However, the decision was stayed pending an appeal filed on Carrie's behalf. Just two weeks before the decision, Dr. Priddy, who had been ailing, died from Hodgkin's disease.[265] He would not live to see the results of the trial. Priddy's successor at the colony was John Bell. Strode petitioned the court to add Bell's name to the case. This request was approved, and the case's name was permanently changed to *Buck v. Bell.*

On June 1, 1925, Irving Whitehead filed an appeal on behalf of Carrie Buck to the Virginia Supreme Court of Appeals, the state's highest court. As with any appeal, there was no opportunity to present any new evidence or challenge the factual record established at the first trial. The court would review only the legal arguments presented on appeal by the respective attorneys. The petition presented by Whitehead simply reasoned that the Amherst County Circuit Court erred in its decision to allow Carrie to be sterilized. The petition also raised three constitutional arguments: (1) that sterilization violated the due process clause of the Fourteenth Amendment, (2) that it violated the equal protection clause of the Fourteenth Amendment and (3) that it would inflict cruel and unusual punishment in violation of the Eighth Amendment.[266] These constitutional arguments were tactfully raised, because if the appeals court rejected them—as it was expected that it would—the case would then be brought before the U.S. Supreme Court.

Each attorney drafted legal briefs in support of their argument. Such briefs would typically contain cogent legal analysis with supporting case law and other exhibits. Unsurprisingly, Whitehead submitted a poorly drafted legal brief just five pages long. Conversely, Strode's brief was forty-four pages long, with numerous citations, and presented a comprehensive legal argument in support of the colony's case. Among Strode's arguments was the critical notion that the state required a sterilization law "to protect the public health and the public safety."[267] In support of this argument, Strode relied on a legal analogy, proffered by Harry Laughlin, between Carrie's case and *Jacobsen v. Massachusetts*, a Supreme Court case that upheld the legality of compulsory vaccinations against smallpox in 1905.[268]

On November 12, 1925, the Virginia Supreme Court of Appeals affirmed the lower court's decision. The court agreed that the sterilization procedure as described by the colony was safe and that undergoing such a procedure was in Carrie's best interest because it would liberate her. The court rejected the constitutional arguments presented by Whitehead. It ruled that the Virginia sterilization statute provided due process, was not discriminatory and was not cruel and unusual punishment. Despite the victory, Strode advised hospitals in the state to hold off from administering sterilizations until the matter was fully and finally heard before the highest court in the nation.

Irving Whitehead remained Carrie's attorney, and his betrayal of her as his client continued unabashed. On December 7, 1925, he joined Aubrey Strode to meet with the colony's board to present an update on the trial. The following passage was noted in the minutes of that meeting:

> *Colonel Aubrey E. Strode and Mr. I.P. Whitehead appeared before the Board and outlined the present status of the sterilization test case and presented conclusive argument for its prosecution through the Supreme Court of the United States, their advice being that this particular case was in admirable shape to go to the court of the last resort, and that we could not hope to have a more favorable situation than this one.*[269]

In the spring of 1927, *Buck v. Bell* finally appeared before the United States Supreme Court. After a string of judicial losses, all on constitutional grounds, this was the moment that eugenicists had long waited for, and with a case as strong as this one—thanks to the assistance of Irving Whitehead, who failed to dutifully serve his client—their anticipation was high. Once again, legal briefs were submitted by both Whitehead and Strode, and as with the previous appeal, Whitehead's brief was weak and limited, while Strode presented a strong and thorough legal analysis. Oral arguments were presented on April 22, 1927, and the justices then deliberated on the case.

On May 2, 1927, the United States Supreme Court issued its ruling. In an 8–1 decision, the court upheld the lower court's ruling that Carrie Buck should be sterilized.[270] The opinion, which was drafted by Justice Oliver Wendell Holmes Jr., was a mere five paragraphs in length. In it, Holmes stated that the Virginia sterilization law was necessary "to prevent our being swamped with incompetence." He then issued perhaps one of the most haunting passages in the history of American jurisprudence when he wrote,

> *It is better for all the world, if instead of waiting to execute degenerate offspring for their crime, or to let them starve for their imbecility, society can prevent those who are manifestly unfit from continuing their kind.... Three generations of imbeciles are enough.*[271]

Much of the decision appeared to reflect the Court's indifference to the plight of Carrie Buck and all those who would later be subjected to forced sterilization. Later in life, Holmes told a friend that *Buck v. Bell* was "one decision that I wrote that gave me pleasure."[272] Although shocking, it should come as no surprise, given the successful infiltration of eugenics in all walks of American life, that such chilling and seemingly mean-spirited language would be offered by the chief justice of the highest court in the United States.

The Court had no interest in rectifying any factual inconsistencies in the record so carefully orchestrated by Strode and Whitehead at the first trial.[273] Nevertheless, there were numerous inconsistencies that could have at least been addressed. Perhaps the greatest of these inconsistencies was that Justice Holmes labeled Carrie an "imbecile," which was different from "middle-grade moron," as the Colony diagnosed her to be, and very different from "feebleminded," as the trial witnesses declared her to be. This suggests that the high court never understood or cared about these supposed medical terms and simply determined that Carrie Buck should be sterilized. Moreover, the reference made by Holmes to "three generations of imbeciles" will always remain in doubt because Vivian was never truly tested. Lastly, Holmes accepted Laughlin's analogy between Carrie's case and the *Jacobsen v. Massachusetts* smallpox vaccination case. In the decision, he wrote, "The principle that sustains compulsory vaccination is broad enough to cover cutting the Fallopian tubes."[274]

On July 3, 1927, just seven weeks after the trial, Carrie's daughter, Vivian, tragically died of an intestinal infection following the measles.[275] Three months later, on October 19, 1927, Carrie Buck was sterilized. She was twenty-one years old. The procedure was performed by Dr. John Bell at the colony's Halsey-Jennings building. Even though the Dobbs family insisted during the trial that they would take Carrie back, they ultimately refused to accept her in their home. On January 23, 1928, Carrie's sister Doris, who was diagnosed by the colony as a "high-grade moron," was also sterilized. She was only sixteen years old.[276]

After an extensive search by Dr. Bell, Mr. A.T. Newbury of Bland, Virginia, agreed to take Carrie into his family home. It was agreed that he would pay Carrie five dollars per month to perform household chores. Finally, on January 1, 1929, Carrie Buck was formally discharged from the colony. On May 14, 1932, the twenty-five-year-old Carrie married a sixty-three-year-old widower named William D. Eagle. Carrie's new husband had four daughters and two sons from a previous marriage. On April 15, 1944, Emma Buck died of bronchial pneumonia at the age of seventy-one.

Left: Vivian Buck, daughter of Carrie Buck, seated on the lap of Alice Dobbs in Charlottesville, Virginia, circa 1924. Just three years later, on July 3, 1927, Vivian Buck died from an intestinal infection after contracting the measles. *Photo courtesy of Arthur Estabrook Papers, M.E. Grenander Special Collections & Archives, University at Albany, SUNY.*

Below: Carrie Buck (*left*) seated next to her mother, Emma Buck, circa 1924. Both women were declared "feebleminded" based on false information. *Photo courtesy of Arthur Estabrook Papers, M.E. Grenander Special Collections & Archives, University at Albany, SUNY.*

The U.S. Supreme Court, also known as the "Taft Court" (after Chief Justice William H. Taft). The Taft Court's term spanned from 1921 to 1930. On May 2, 1927, the Court upheld a request by the State of Virginia to forcefully sterilize nineteen-year-old Carrie Buck based on a eugenic diagnosis. The 8–1 decision, a mere five paragraphs in length, was written by Justice Oliver Wendell Holmes Jr. (*seated, second from right*), who later told a colleague that "it was one decision that I wrote that gave me pleasure." The 1927 decision (*Buck v. Bell*, 274 U.S. 200) paved the way for thousands of eugenic sterilizations across the United States and was cited by doctors of the Nazi regime during their trial in Nuremberg after the Second World War. *Photo courtesy of the Library of Congress.*

Carrie Buck went on to live a relatively simple life until her death on January 28, 1983. However, she never recovered from being sterilized against her will. During an interview later in life, she insisted that she wanted to have more children. She stated, "Oh yeah, I was angry. They done me wrong. They done us all wrong."[277]

Buck v. Bell was the legal victory that eugenicists had long coveted. It confirmed that compulsory sterilization of patients at state hospitals throughout the nation, based on dubious eugenic diagnoses, could lawfully proceed. This inevitably led to the sterilization of thousands of people throughout the country. It took a great deal of time and effort to get to that point, but none of it would have reached fruition without the tireless efforts of the Eugenics Record Office and more specifically Harry Laughlin himself, who spent years perfecting and promoting model statutes that states could use to make eugenic sterilization into law, including the Virginia sterilization law. When the ideal test case finally arose, a case that would reach the United States Supreme Court, Laughlin did everything in his power to assist. He delivered sworn, albeit factually incorrect, testimony that was admitted into the trial record. He coached and dispatched Arthur Estabrook to provide critical expert witness testimony. Laughlin even conceived of the legal analogy between eugenic sterilization and compulsory vaccination, an argument accepted and repeated by Justice Oliver Wendell Holmes Jr. in the Court's decision.

Buck v. Bell has never been reversed. It is an enduring legacy left by the Eugenics Record Office and a direct byproduct of the ERO's work. In the wake of the decision, the number of sterilizations across the country began to grow exponentially. While many states later repealed their eugenic sterilization laws, there continued to be a sharp increase in coerced sterilizations toward the end of the 1960s, all of which started with the eugenic sterilization at the beginning of the twentieth century. *Buck v. Bell* was so impactful that, in November 1946, the Nazi doctors who were on trial in Nuremberg cited it in defense of the atrocities they had committed during the Second World War.

Chapter 11

KILLING IN THE NAME OF...

From an historical point of view, the first method which presents itself is execution.…Its value in keeping up the standard of the race should not be underestimated.
—Paul Popenoe, 1918

Beginning in 1911, members of the eugenic section of the American Breeders Association contemplated and urged several dangerous measures to deal with those who were deemed unfit for society. At the direction of the board, Harry Laughlin drafted his infamous report in 1914, which proposed ten remedies for "cutting off the defective germ-plasm."[278] These remedies included restrictive marriage laws, life segregation and sterilization. No. 8 on the list was euthanasia. By no means was this list intended to be rhetorical. It was both a mandate and blueprint supported by the trappings of cutting-edge eugenic science and fully intended to be a catalyst for a national and global campaign.

For this proposed remedy of euthanasia, Laughlin praised the courage and sacrifice of the mothers of ancient Sparta whose business was to "grow soldiers for the State" to die in battle.[279] Although the committee ultimately opted to focus on the "prevention of defectives rather than destroying them before birth, or in infancy, or in the later periods of life," the fact that euthanasia was contemplated in the United States on a mass scale is quite frightening.[280]

Discourse on implementing mass euthanasia programs was popular among many eugenicists nationwide. The first discussions about the use of so-called lethal chambers to kill mass amounts of people arose during the Victorian era, when England began a campaign to euthanize stray dogs and cats.[281] In 1900, American physician and eugenicist W. Duncan McKim's *Heredity and Human Progress* was published. In it, McKim wrote: "Heredity is the fundamental cause of human wretchedness....The surest, the simplest, the kindest, and most humane means for preventing reproduction among those whom we deem unworthy of this high privilege [reproduction], is a gentle, painless death." He added, "In carbonic acid gas, we have an agent which would instantaneously fulfill the need."[282]

In 1904, E.R. Johnstone, superintendent of the New Jersey Vineland Training School, publicly spoke about the elimination of the so-called feebleminded. Paul Popenoe, a leading eugenicist from California, also stated: "From an historical point of view, the first method which presents itself is execution....Its value in keeping up the standard of the race should not be underestimated."[283] Madison Grant, author of *The Passing of the Great Race* and president of both the Eugenics Research Association and the American Eugenics Society, openly advocated for the mass elimination of defectives, as did many others. At the time, the focus of American eugenicists' work involved the segregation and sterilization of those they determined to be hereditarily defective. Thus, much of the discussion surrounding euthanasia was largely theoretical. However, that all changed in 1915, when news broke about physicians at a Chicago hospital who purposefully denied treatment to newborn child, who was left to die.

On November 12, 1915, a woman named Anna Bollinger gave birth to a child with a deformity and extreme intestinal and rectal abnormalities. Hospital chief of staff Dr. Harry Haiselden was summoned to consult with his colleagues, and due to the child's medical condition, it was decided that the child was simply not worth saving. As a result, treatment was purposefully withheld until the child succumbed. The deliberate denial of medical attention to a newborn child was likened to eugenic euthanasia, and news of the death soon gripped the nation.

An inquest was conducted, and Dr. Haiselden openly declared that allowing newborn children with birth defects to die was not uncommon at the hospital. In reaffirming his decision, he stated, "I should have been guilty of a graver crime if I had saved this child's life. My crime would have been keeping in existence one of nature's cruelest blunders."[284] Although the doctors who testified believed that the child had at least a one in three chance of survival, it was determined that Dr. Haiselden broke no law and was within his professional rights to refuse to provide treatment.

After the inquest, Dr. Haiselden embarked on a public tour promoting his eugenic ideals and openly declared that, at times, he injected children with opiates to hasten their deaths. In a six-part report published by the *Chicago American*, he justified the killings by linking eugenic euthanasia to the deplorable conditions at the Illinois Institution for the Feebleminded, where windows without screens were purposefully left open so that patients would face drafts and swarms of parasitic insects. Haiselden also alleged that patients at that facility were purposefully given milk from cows that were infected with tuberculosis. Studies conducted at this facility appeared to verify Haiselden's claims, as the death rate for some patient groups was as high as 30 percent, mostly by tuberculosis, and in 1915, the overall death rate for new patients at this and other similar facilities increased from 4.2 percent to 10 percent.

Many at the time were already speculating that homes for the feebleminded were akin to lethal chambers. Haiselden confirmed these suspicions with his public statements and dubbed them "slaughterhouses."[285] Moreover, many of the staff members at the Illinois Institution for the Feebleminded were known to staunchly support eugenics. Thus, it would not be surprising if they practiced eugenic measures. Furthermore, many of the patients at this facility were institutionalized by Judge Harry Olson of the Municipal Court of Chicago, a known eugenicist who later served as president of the Eugenics Research Association and publisher of Harry Laughlin's book *Eugenic Sterilization in the United States*.[286]

Judge Harry Olsen sat on the bench in Chicago, Illinois. He was also a staunch eugenicist who served as president of the Eugenics Research Association and helped publish Harry Lauglin's book *Eugenic Sterilization in the United States*. Judge Olsen condemned many patients to be institutionalized at the Illinois Institution for the Feebleminded. *Photo courtesy of the American Philosophical Society.*

Eugenicists nationwide hailed Dr. Harry Haiselden as a hero. They believed that his refusal to provide medical care and instead allow a deformed child to die was a positive step toward the proposed eugenical remedy of mass euthanasia. In 1917, he was cast in the film *The Black Stork*, a fictionalized story about a couple who is counseled by Haiselden not to have a child because it will likely be defective. The woman eventually gives birth to the child, but she then allows it to die, and when it does, its apparition is seen levitating into the arms of Jesus Christ. *The Black Stork* was a hugely successful eugenic propaganda film that was played regularly in theaters throughout the country for more than a decade.

Regardless of the means used, the killing of people arbitrarily deemed to be hereditarily defective was and always has been the gravest threat presented by eugenics. Dating to ancient Greece, medical professionals in the Western world have been required to take the Hippocratic Oath. This set of medical ethics requires those who practice medicine to exercise proper judgment in the care of the patients they treat.[287] Yet eugenics was so prominent that it enabled and emboldened medical professionals like Harry Haiselden and others to murder patients through the refusal to provide lifesaving treatment or by directly or indirectly infecting patients with tuberculosis and other dangerous diseases. Sadly, these murderous acts were performed under the banner of eugenics, which was later discredited as a science, and those who perpetrated them were never brought to justice.

Chapter 12

GLOBAL REACH

The Eugenics Record Office and the Eugenics Research Association congratulate the German people on the establishment of their new Institute for the Biology of Heredity and Race Hygiene.... We shall all be glad indeed to keep in touch with you in the development of eugenics in our respective countries.
—Harry Laughlin, 1936

American eugenicists like Charles Davenport believed that a perfect human race could be achieved if mankind exercised better breeding while simultaneously eliminating those with inferior hereditary traits. Obviously, international cooperation was required to realize this goal. So even as they were securing increasing levels of institutionalization and sterilization of people throughout the nation, along with a successful campaign for strict limitations on immigration by so-called inferior races from other countries, officials at the ERO continually worked to expand its global network to ensure worldwide support and cooperation for eugenics.

In 1912, the first International Eugenics Congress was held in London. Led by a U.S. delegation, Germany, Belgium, Italy and France all joined the conference to discuss their eugenic ideals. Harry Laughlin's eugenics research was circulated at the conference and ultimately published by the Carnegie Institute.[288] Approximately a dozen of the members in attendance formed the International Eugenic Congress, which met for the first time a year later in Paris. They scheduled the second International Eugenics Congress for 1915, but the event had to be placed on hold when Germany invaded Belgium in August 1914, effectively starting World War I.

The second International Eugenics Congress was rescheduled for September 1921 and was held at the American Museum of Natural History in New York. This international event was carefully directed by Charles Davenport and Harry Laughlin, while Harry Osborn, an American eugenicist, was named the event's president. Other attendees of the American delegation included Mary Harriman, Alexander Graham Bell and Madison Grant, a virulent racist and author of *The Passing of the Great Race*, a book that was later praised by Adolf Hitler.[289] The event was well funded by generous donations from Mary Harriman and the Carnegie Institute. Attendees were treated to more than 120 eugenics-based exhibits, which were divided into four sections titled Comparative Heredity, the Human Family, Racial Differences and Eugenics and the State. Nearly all the more than fifty scientific papers that were submitted were produced by American eugenicists. A special exhibit displaying all the U.S. sterilization statutes drew particularly high praise from those in attendance.

Eager to continue spreading American eugenics, Harry Laughlin was dispatched on a much-anticipated European tour. In September 1923, he attended the Permanent International Commission to peddle the so-called Ultimate Program. Devised by the American Eugenics Society, the four-part plan called for each nation to adopt eugenic research, education, administrative measures and conservative legislation within its borders.[290] Davenport and Laughlin oversaw the commission, and three delegates from each participating country were empaneled to assist. Legislative, scientific and political resolutions were drafted, and their terms were binding on the members of the commission.

By 1925, the Permanent International Commission had grown to include prominent individuals from various medical and scientific fields, along with eugenic societies and institutions. The group was renamed the International Federation of Eugenic Organizations (IFEO).[291] Many members of the group frequently traveled to the ERO in Cold Spring Harbor to receive training and attend meetings and conferences. While the United States remained the capital of the eugenics world, other countries began to develop their own organizations furthering the eugenic crusade. The Belgian Eugenical Society was organized in 1919 under the leadership of Dr. Albert Govaerts. A personal colleague of Harry Laughlin, Govaerts attended the second International Eugenics Congress and later studied at the ERO, with much of his work focused on tuberculosis. Under his leadership, the Belgian Eugenical Society established courses at the University of Brussels, and much like at the ERO, it formed a National Office of Eugenics to train field-workers.

In sweeping fashion, the eugenic practices that were devised and perfected at Cold Spring Harbor directly springboarded to the capitals of other nations. Driven by increased immigration in the late nineteenth century, primarily of people from Asia and Europe, eugenics began to take hold in Canada. In 1905, Ontario created its first census of feebleminded individuals. A few years later, the British-American Medical Association in Canada was studying the sterilization laws already enacted in the United States. In 1928, Alberta's legislature passed Canada's first eugenic sterilization law, which targeted mental defectives who "risk…multiplication of [their] evil by transmission of [their] disability to progeny."[292] Within the first nine years of its formation, the Alberta Eugenics Board authorized the sterilization of approximately four hundred people.

In 1910, under the leadership of psychiatrist Dr. Auguste Forel, Switzerland also adopted eugenic policies. A wealthy industrialist named Julius Klaus became a staunch advocate who, on his death in 1920, bequeathed more than one million francs (an amount equal to $4.4 million in today's money) to establish a fund for eugenics in the country. Well-funded organizations such as the Julius Klaus Foundation for Heredity Research, Social Anthropology and Racial Hygiene, along with the Institute for Race Biology, were established. Swiss eugenics targeted certain ethnic groups and the study of sexual behavior among women. The first sterilization law in the country was passed in 1928, and while exact figures remain unknown, it is estimated that approximately 90 percent of the procedures were performed on women.

In Denmark, eugenics was largely spearheaded by two men: August Wimmer, a psychiatrist at the University of Copenhagen, and Soren Hansen, president of the Danish Anthropological Committee. Both men were devoted to the work of Charles Davenport. One physician from Denmark traveled to the Vineland Training School for Feeble-Minded Boys and Girls to study under Henry Goddard, whose eugenics work and revision of the Binet-Simon test became the leading standard followed by Danish eugenical publications.[293] A sterilization law in Denmark was enacted in 1929 with the direct assistance of Harry Laughlin and Harry Olson, a staunch eugenicist and judge from Chicago. Soon thereafter, the Rockefeller Foundation began to support eugenics research in Denmark. A grant was awarded to Dr. Tage Kemp, a leading eugenic scientist, who in 1932 traveled to the ERO in Cold Spring Harbor to further his studies.

In Norway, Jon Alfred Mjoen, a scientist and close ally of Charles Davenport, was an early and fervent supporter of eugenics. That country

Staff and Members - Dept. of Genetics, C. I. of W., May 12, 1934.

1. Alice Gould Laanes
2. Ruth Millar
3. Ethel Burtch
4. Margaret Martin
5. Ethel I. Hunt
6. Martha Taylor
7. Margaret Hoover
8. Margaret Finley
9. Lillian Frink
10. Ingar Andersen
11. Mary J. Holmes
12. Dr. Dorothy Bergner
13. Lincoln Cartledge
14. Dr. Robert Bates
15. Dr. M. Demerec
16. Dr. Oscar Riddle
17. Dr. A. F. Blakeslee
18. President, John C. Merriam
19. Director, Charles B. Davenport
20. Walter M. Gilbert
21. Dr. Harry H. Laughlin
22. Dr. Clarence Moran
23. Dr. James Potter
24. Dr. E. C. McDowell
25. Dr. Morris Steggerda
26. Merriam North
27. Gwendolyn Smith
28. Elizabeth McKee
29. J. D. McGlohon
30. Madeleine Wilkins
31. Gabriel A. Lebedeff
32. Margaret Kaylor Kuntz
33. Eunice White
34. Mrs. Isabel Griffin
35. Mrs. Harriet Smart
36. Jennie Schultz
37. Mrs. L. Carteledge
38. Amos Avery
39. Mrs. Amos Avery
40. Catherine Carley
41. Edith Harrigan
42. Alice Hellmer
43. Mrs. Hilda Wüllen
44. George W. Macarthur
45. William Murray
46. Leslie E. Peckham
47. Dr. Benjamin Speicher
48. Theophil Laanes
49. Floyd Matson
50. James J. Banta
51. E. L. Lahr
52. J. P. Schooley
53. George A. Smith
54. William Drager
55. Louis Stillwell
56. William Schneider
57. Paul O. Holm
58. William Fagan
59. Peter Campbell
60. Edward Burns
61. Mrs. John Buccuris
62. John Buccuris
63. Stanley Brooks
64. John N. Johnson
65. Clifford Valentine
66. Harry White
67. Dominico Sepe.

passed its first sterilization law in 1934, which remained in place until 1977, when sterilization was converted to a voluntary procedure. Approximately forty-one thousand surgical procedures were performed, nearly 75 percent of them on women. Sweden opened the State Institute of Race-Biology in 1922, which was entirely dedicated to eugenics.[294] In 1934, the country passed its first sterilization law. Initially targeting patients with "mental illness, feeble-mindedness or other mental defects," the law expanded to include those with "an anti-social way of life." Ultimately, sixty-three thousand sterilizations were conducted in Sweden, mostly on women.

Eugenics continued to spread throughout Europe to countries like Finland, Hungary, France, Romania, Italy and others. All these programs benefited directly from funding by the Rockefeller Foundation and the Carnegie Institute. By the time of the third International Eugenics Congress, which took place in New York in 1932, eugenics was fully legitimized as a true science and recognized internationally.

While many countries around the world established programs for immigration limitations, mass segregation and sterilization of the eugenically unfit, it was unknown at the time if other, far more draconian, measures would be adopted. The world would soon find out once Germany entered the eugenic arena.

Opposite: Staff members of the Carnegie Institute Department of Genetics on May 12, 1934. President John C. Merriam is seated at the center (no. 18). To his left is Director Charles Davenport (no. 20) and Harry Laughlin (no. 21). *Photos courtesy of the Truman State University, Pickler Memorial Library, Special Collections and Museums.*

Chapter 13

EUGENICS AND THE THIRD REICH

The next round in the thousand-year fight for the life of the Nordic race will probably be fought in America.
—German eugenicist Fritz Lenz, 1923

The distance between Cold Spring Harbor, New York, and Berlin, Germany, is approximately 3,928 miles. During the early twentieth century, the typical mode of transportation between these destinations was by boat, which required weeks of travel. Of course, there was no internet or social media at that time, so immediately communicating and influencing an audience across the globe was far more difficult. This meant that eugenicists had to work much harder to spread their ideals, and officials at the ERO were more than eager to perform this work. This chapter explores the direct connection between the leaders of the Eugenic Records Office and their eugenic counterparts in Germany, including the ERO's direct relationship with Adolf Hitler's Nazi regime and his menacing team of scientists and doctors who committed numerous atrocities under the banner of eugenics.

For much of the first half of the twentieth century, the United States led the world in eugenics, and the Eugenics Record Office was the very epicenter of this movement. However, German scientists and physicians became increasingly interested in the pseudoscience. One such physician was Alfred Ploetz, who toured the United States and studied the writings of American eugenicists who were interested in breeding better human beings. Ploetz opened a medical practice in Springfield, Massachusetts, and began to breed chickens.[295] As his interest in eugenics grew, he relocated to Connecticut and compiled the genealogies of 325 local families. Ploetz became a disciple of the hygienic movement, which sought to eradicate germs and disease and eventually adopted eugenics-based racial theories. By the late 1880s, Ploetz coined the term *Rassenhygiene*, or "racial hygiene." Eventually, Ploetz emerged as one of the early leaders of eugenics in Germany, and in 1905, he founded the Society for Racial Hygiene in Berlin. In 1912, he was recruited as a vice president of the first International Eugenics Congress in London and was invited back a year later to form the Permanent International Eugenics Committee.[296]

At its core, eugenics is premised on the belief that only superior races have the best blood. This and similar ideals drew great interest from German race hygienists, many of whom maintained strong relationships with Charles Davenport. One such relationship existed with Eugen Fisher, a German anthropologist who began corresponding with Davenport as soon as the Station for Experimental Evolution at Cold Spring Harbor opened in 1904. At the time, both men were conducting research on eye color and hair quality, but their mutual interests later turned to miscegenation, a term used to describe the reproduction of people from different ethnic groups.[297]

Over the years, the German interest in American eugenics continued to grow. In 1913, an Austro-Hungarian vice counsel named Geza von Hoffmann traveled throughout the United States to study American eugenics. He authored a book titled *Racial Hygiene in the United States*, which detailed American laws on sterilization and marriage restrictions. Hoffmann wrote, "Galton's dream that racial hygiene should become the religion of the future, is being realized in America....America wants to breed a new superior race."[298] Hoffmann deeply admired the fact that in the United States, many of those with inferior traits were being confined to institutions. He bitterly complained about the lack of race-based immigration barriers in his country and particularly lamented over the increasing number of mental defectives freely roaming the streets of Germany.

The overwhelming source of Hoffmann's eugenic information was received directly from the ERO and specifically from Harry Laughlin. In a letter to Laughlin dated May 26, 1914, Hoffmann wrote, "The far-reaching proposal of sterilizing one tenth of the population impressed me very much."[299] Later that year, Hoffmann authored an article titled "Eugenics in Germany," which was published in the *Journal of Heredity*. Although the practice of eugenics in Germany largely stalled during the First World War, the ties between Hoffmann and the ERO continued to strengthen. Once the war ended in June 1919, the American-German eugenic partnership resumed in earnest.

In June 1920, Harry Laughlin prepared a pro-German speech for the ninth annual meeting of the Eugenics Research Association at Cold Spring Harbor. In the text of the speech, he analyzed Germany's newly imposed constitution and concluded, "From what the world knows of Germanic traits, we logically concede that she will live up to her instincts of race conservation."[300] While he never actually delivered that speech, Laughlin published it in *Eugenical News*, which was widely circulated throughout the eugenics world.

In 1929, Charles Davenport initiated a global campaign to eradicate the mixing of races. He had previously embarked on a two-year study on the island nation of Jamaica of "pure-blooded negroes, as found in the western hemisphere…and of whites, as found in the same places with especial reference to inheritance of the differential traits in mulatto offspring."[301] The ERO paid $10,000 to defray the cost of the study. Within two years, Davenport had studied the family backgrounds of 370 individuals, largely inmates at a nearby prison and citizens in the capital city of Kingston.

Davenport was eager to expand his work on race mixing with his international colleagues. Under his signature as president of the International Federation of Eugenic Organizations (IFEO), he sent numerous letters to his international contacts stating: "The committee on race crossing of the Federation is seeking to plot the lines, or areas, where race crossing between dissimilar, more or less pure races is now occurring or has been occurring during the last two generations."[302] Both Davenport and Eugen Fischer coauthored a questionnaire and had it translated into English, Spanish and German before distributing it globally.

On September 27, 1929, Davenport and Fischer convened a special meeting in Rome, Italy. The goal of the meeting was to create a global atlas of the defective populations on every continent and discuss eugenic measures to deal with these groups. Those in attendance included leading eugenicists from the United States, Sweden, Norway, Holland, Italy, England and Germany.[303] The group discussed their findings and examined numerous maps and surveys.

One of the key objectives of the meeting was to formulate a measure to prevent members of the eugenically unfit populations from marrying outside of their own groups. Jon Alfred Mjoen of Norway identified those in his country who were suffering from tuberculosis, while the Dutch representative focused on the "mixed breeds" of the Java islands. Captain

George Pitt-Rivers from England suggested that anthropologists should compile ethnographic statistics in each country, and Davenport discussed the intelligence testing administered to soldiers in the U.S. Army, along with tuberculosis rates in Virginia. A consensus was reached among all in attendance that the permanent incarceration of paupers, mental defectives, criminals, alcoholics and other so-called inferiors was required.

AS GERMANY BEGAN TO match the United States in eugenic research, both countries agreed to form an equal partnership to freely exchange information and ideas on eugenics. The *Eugenical News* provided an amplifying voice for German eugenics. In 1926, it featured an article written by Fritz Lenz, "Are the Gifted Families in America Maintaining Themselves?" In it, Lenz warned:

> *The dying out of the gifted families…of the North America Union* [United States] *proceeds not less rapidly; and also among us in Europe….I think one ought not to look at the collapse of the best element of the race without action.*[304]

A key focus of German eugenics was the racist belief that Jews were eugenically undesirable. Even though this theory lacked any scientific support, many of the American eugenicists adopted the same belief, and such rhetoric was routinely amplified by the *Eugenical News*. In May 1927, the *Eugenical News* also reported on Germany's introduction of a "race biological index." This was intended to rate certain racial groups over others and reiterate German warnings of an "eruption" of colored races throughout Europe.

The *Eugenical News* also promoted several racist German articles. In November 1925, the *Eugenical News* published a German article protesting interracial marriages. A portion of that article stated, "Their evil consequences…are pointed out [and]…are commoner among Jews and royalty than elsewhere in the population." In December 1927, the *Eugenical News* promoted another German article that discussed the social biology and hygiene of Jews, and in an April 1929 edition, the publication provided a glowing review of the book *Noses and Ears*, which discussed the convex nose shape of Jewish people.

Julius Lehmann was a German publisher and close confidant of Adolf Hitler who enjoyed a great deal of favorable reporting in the *Eugenical News*.

Lehmann produced a series of racist trading cards depicting the racial profiles of peoples from the Tamils of India to the Bashkirs of the Ural Mountains. These cards were proudly reported on by the *Eugenical News*. In April 1924, the *Eugenical News* also offered a glowing review of a so-called racial pride book published by Lehmann, which, among other topics, reviewed the history and societal role of Jews in Germany.

As the ERO was eagerly providing an amplifying voice for German bigotry and vitriol, American money helped fund it. German scientists at the Kaiser Wilhelm Institutes, who began to make great strides in eugenics research and biomedicine, received generous support from U.S. organizations like the Rockefeller Foundation, including financial, administrative and scientific support for a wide array of the organizations that were formed under the Kaiser Wilhelm umbrella, including the Kaiser Wilhelm Institute for Psychiatry; the Institute for Anthropology, Human Heredity and Eugenics; and the Institute for Brain Research. Some of these organizations were directly responsible for the medical atrocities ultimately committed by the Nazi regime.

By the end of 1922, the Rockefeller Foundation had awarded nearly 200 fellowships totaling $65,000 to these German organizations. The following year, 262 fellowships were awarded for a total of $135,000. By 1926, the Rockefeller Foundation had donated a total of $410,000 to hundreds of German researchers, an amount equal to $4 million in today's money.[305] Officials at the Rockefeller Foundation were also fascinated with the field of psychiatry, and in May 1926, the foundation donated $250,000 to the Kaiser Wilhelm Institute of Psychiatry, followed by an additional $75,000 a year later.

Ernst Rudin was considered by many to be the most prominent race hygienist in Germany. Rudin was one of the founders of the Kaiser Wilhelm Institute of Psychiatry, and his work focused on the need to stop what he felt was the dangerous breeding of the eugenically unfit. Like Davenport at the ERO, Rudin began to assemble family pedigrees gleaned from records collected at prisons, churches, insane asylums and hospitals. He also forged strong ties with government officials and with the administrators of these facilities.

Rudin's work was lauded in the United States. In May 1922, the *Journal of Heredity* published an interview with him on the inheritance of mental defects. In 1924, the *Eugenical News* reported on his efforts to construct family histories and added that his studies of "inheritance of mental disorders are the most thorough that are being undertaken anywhere. It is hoped that they will long be continued and expanded."[306] The *Journal of the American Medical Association* also published a glowing review of Rudin's work on heredity and mental disease.

In 1927, the Kaiser Wilhelm Institute formed a eugenic section called the Institute for Anthropology, Human Heredity and Eugenics. Eugen Fischer was named its director. A grand opening was held in Berlin in connection with the fifth International Congress on Genetics. Nearly one thousand delegates were in attendance, including Fischer's good friend Charles Davenport, who served as chairman of the human eugenics program and honorary president of the congress. By the late 1920s, under Davenport's leadership, American eugenicists had established numerous joint projects with their German counterparts intended to spread eugenics worldwide.

Davenport's project in Jamaica featured for the first time the use of IBM's Hollerith data-processing machines. The personal information of subjects, including their genetic traits, was fed into these machines at the rate of twenty-five thousand cards per hour. These high-speed tabulators kept track of a wide array of eugenic subjects, and a system was developed to track and report certain racial characteristics. Just five years later, IBM's president, Thomas J. Watson, enhanced this high-speed system, which was adapted to automate the race warfare and persecution of Jews by the Hitler regime.[307] A decade later, the Nazi Schutzstaffel (known as the SS) Race Office utilized a similar system to categorize people based on physical attributes in a meticulous column-by-column fashion.

In the 1930s, the Rockefeller Foundation continued to provide generous funding for eugenics in Germany. This included a donation of $125,000 to Fischer's German anthropological study, which by then was largely directed at dealing with Jews in Germany.[308] In December 1930, *Eugenical News* published Rudin's report titled "Hereditary Transmission of Mental Diseases." In the report, he declared:

> *Humanity demands that we take care of all that are diseased—of the hereditarily diseased too—according to our best knowledge and power; it demands that we try to cure them from their personal illnesses. But there is no cure for the hereditary dispositions themselves. In its own interest, consequently, and with due respect to the laws of nature, humanity must not go so far as to permit a human being to transmit his diseased hereditary dispositions to his offspring. In other words: Humanity itself calls out an energetic halt to the propagation of the bearer of diseased hereditary dispositions.*[309]

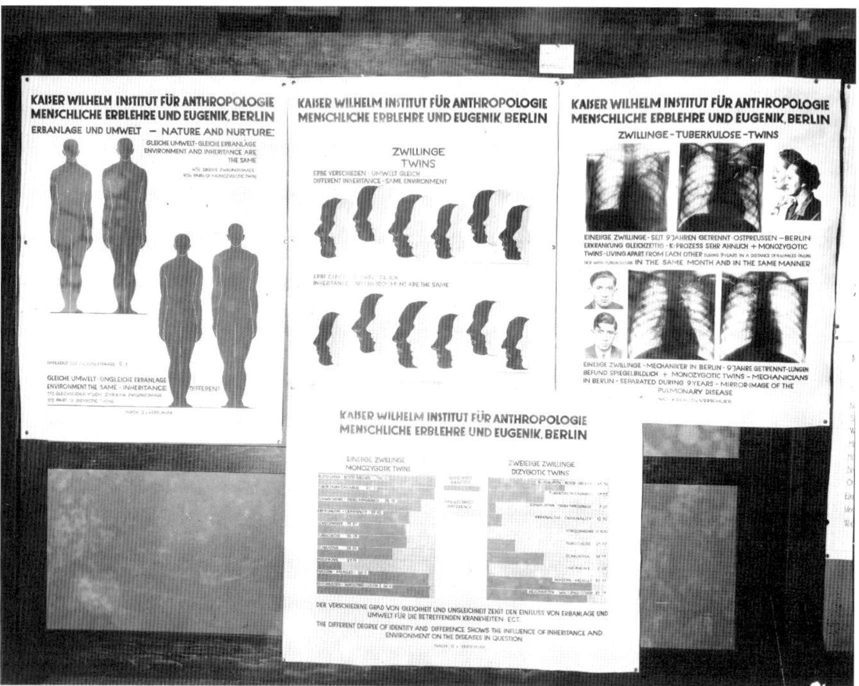

A German eugenics exhibit on the study of twins created by the Kaiser Wilhelm Institute, on display at the Third International Eugenics Congress in 1932. *Photo courtesy of the Truman State University, Pickler Memorial Library, Special Collections and Museums.*

Despite the dangerous rhetoric and research, funding for the German eugenic programs continued. In 1931, the Rockefeller Foundation approved an additional ten-year grant to Rudin's Institute for Psychiatry totaling $89,000.

ON NOVEMBER 3, 1923, Adolf Hitler led the infamous Beer Hall Putsch in Munich. The attempt to seize power was quickly suppressed, and he was sent to prison for five years. While in prison, he dictated his notorious *Mein Kampf* to Rudolf Hess, but he also read many books of great interest to him. One of those books was *The Foundation of Human Heredity and Race Hygiene*, which was published in 1921. The two-volume set was written by Erwin Baur, Fritz Lenz and Eugen Fischer, all of whom were eugenicists and directly linked to Charles Davenport.[310]

Erwin Baur was a leading German scientist. At the urging of Charles Davenport, Baur joined the International Eugenics Commission. On

November 20, 1920, Baur wrote to Davenport to request information on U.S. marriage restriction laws and other material gathered by the Eugenics Record Office. Fritz Lenz had long been a staunch admirer of the American eugenics program. While he already had established a relationship with Paul Popenoe, a leading eugenicist from California, he also often requested publications from Davenport at the ERO. Lenz later predicted that "the next round in the thousand-year fight for the life of the Nordic race will probably be fought in America."[311] The third author whose work inspired Hitler while he was in prison was Eugen Fisher, who had been collaborating with Charles Davenport since 1904.

The Foundation of Human Heredity and Race Hygiene by Baur, Lenz and Fischer received high praise from the ERO, and Davenport promised to publish a glowing review in the *Eugenical News* and the *Journal of Heredity*. The book's bibliography included a wide array of American publications, many of which emanated from the ERO. It was published by Julius Lehmann, a staunch promoter of eugenics and a close confidant of Hitler and coconspirator of the Beer Hall Putsch in 1923.

Many of Hitler's beliefs were directly inspired by the eugenics books he read while he was in prison. He also admired the policies of the American eugenics program, including the efforts that led to the passage of strict immigration laws in the United States based on the quest to preserve the purity of the Nordic race. Hitler praised Darwinian principles and condemned charitable efforts. Hitler once told a Nazi official:

> *Now that we know the laws of heredity, it is possible to a large extent to prevent unhealthy and severely handicapped beings from coming into the world. I have studied with great interest the laws of several American states concerning the prevention of reproduction by people whose progeny would, in all probability, be of no value or be injurious to the racial stock...But the possibility of excess and error is still no proof of the incorrectness of these laws. It only exhorts us to the greatest possible conscientiousness....It seems to me the ultimate in hypocrisy and inner untruth if these same people* [social critics]*—and it is them, in the main—call the sterilization of those who are severely handicapped physically and morally and of those who are genuinely criminal a sin against God. I despise this sanctimoniousness.*[312]

Despite his full embrace of eugenics, Hitler privately feared that even he would not meet the high expectations established by the science. He once told a fellow Nazi official,

> *I was shown a questionnaire drawn up by the Ministry of the Interior, which it was proposed to put to people whom it was deemed desirable to sterilize. At least three-quarters of the questions asked would have defeated my own good mother. If this system had been introduced before my birth, I am pretty sure I should have never been born at all.*[313]

Of course, Hitler would suppress his own personal doubts, and by 1931, his calls for the fascist persecution of Jews and others were continuing to grow louder, and public anti-Semitism was sweeping across Germany. In 1932, *Eugenical News* published an essay, titled "Hitler and Racial Pride," that glorified the rising leader. The term *Aryan* began to grow synonymous with the traditional *Nordic*. The *New York Times* reported that the Hitlerites held the Nordic race to be "the finest flower on the tree of humanity" and that it must be "bred and secured…to the criteria of race hygiene and eugenics."[314]

On January 30, 1933, Germany underwent an inconclusive election and Hitler seized power. The depths of his virulent racism and anti-Semitism can never truly be explained, but it can be stated with certainty that eugenics presented Hitler with a legitimate and globally accepted science to support his sinister plans.

In July 1933, Germany enacted the "Law for the Prevention of Defective Progeny," the first eugenic sterilization law in the country. Ernst Rudin assisted in the drafting of the law, and it was soon announced that as many as 400,000 Germans would ultimately be subjected to the sterilization procedure.[315] The procedure became known as the Hitlerschnitte, or "Hitler's Cut." Later that summer, the *Eugenical News* proudly published the Nazi sterilization law, which mimicked similar laws passed in the United States years earlier. The *Journal of the American Medical Association* reported on the law and accepted the unchallenged Nazi eugenic data to support the law as fact. During 1934, the Third Reich sterilized approximately 56,000 individuals, nearly 1 out of every 1,200 Germans.[316] The law also established approximately two hundred eugenic courts and mandated anyone suspected of having a genetic defect to be reported to the authorities. Local newspapers in the United States covered this story, reporting that German "incurables" as young as ten and as old as fifty would be subjected to sterilization.[317]

On January 8, 1934, IBM opened a large factory in Berlin to manufacture Hollerith machines to help tabulate Hitler's first census. Mathematical formulas and high-speed data processing along with medical records became a critical element of Jewish persecution in Nazi Germany. One year later,

German eugenicists began to formulate definitions of Jewishness. Hitler insisted that Jews of all degrees be identified, including those with at least one drop of Jewish blood. This was reminiscent of Virginia's Racial Integrity Act of 1924. The Law for the Protection of German Blood and German Honor and the Reich Citizenship Law were enacted, which deprived Jews of German citizenship. These laws applied to all Jewish people based on mathematical formulas fed into the Hollerith system that could identify Jewish descent to ratios of one half, one quarter, one eighth or one sixteenth. Though the technology was new, the methodology was fully inspired by the family pedigree system created at Cold Spring Harbor more than two decades earlier.

Over time, the world learned of Germany's prosecution of Jews, and global condemnation began to grow. Undeterred, American eugenicists attempted to place a positive spin on the actions of Nazi eugenicists. In 1937, the ERO became a distributor of a two-reel Nazi eugenic propaganda film titled *Erbkrank* (in English, *The Hereditary Diseased*). Even as Jews were facing persecution in Germany, Laughlin continued for more than a year to loan the film to high schools in New York and New Jersey, to welfare workers in Connecticut and to the Society for the Prevention of Blindness.[318] On November 9–10, 1938, Nazis in Germany torched Jewish synagogues, schools and businesses. More than one hundred Jews were murdered in the event known as Kristallnacht, or the Night of Broken Glass. On September 1, 1939, Germany led its blitzkrieg into Poland, and the world was soon plunged into the Second World War.

Adolf Hitler was unquestionably and directly inspired by the American eugenics program that was spearheaded by the Eugenics Record Office. However, Germany's eugenics program was forged more directly by relationships between Charles Davenport, Harry Laughlin and leading doctors in Germany who would later carry out ghastly experiments in the name of eugenics. One of these doctors was Otmar Freiherr von Verschuer, a virulent racist and nationalist who had long argued that fighting Jews was integral to Germany's eugenic battle. His fervor was recognized by Hitler, who said he was "the first statesman to recognize hereditary biology and race hygiene."[319]

In 1925, Verschuer was selected to serve as the secretary of the Society for Race Hygiene in Germany. American eugenicists publicly applauded his appointment, including Paul Popenoe, C.H. Danforth (an anatomist from

Stanford University) and Henry Goddard. The ERO hailed his appointment as well in *Eugenical News*, along with the numerous articles he had written on eugenics. In January 1934, the *Journal of the American Medical Association* cited Verschuer's work: a presentation he gave to the German Congress of Gynecology. Later that year, Verschuer launched *The Genetic Doctor*, a medical journal in which he urged all doctors in Germany to become "genetic doctors." The ERO published a story about the 1935 opening of the Institute for Hereditary Biology and Racial Hygiene in Frankfurt, Germany, and added, "*Eugenical News* extends best wishes to Dr. O. Freiherr Verschuer for the success of his work in his new and favorable environment."[320] The article hailed Verschuer's eugenic mission as one for all mankind.[321]

Harry Laughlin regularly exchanged correspondence with Verschuer. In one letter, he wrote, "The Eugenics Record Office and the Eugenics Research Association congratulate the German people on the establishment of their new Institute for the Biology of Heredity and Race Hygiene.…We shall all be glad indeed to keep in touch with you in the development of eugenics in our respective countries."[322] Verschuer responded to Laughlin, "You have not only given me pleasure, but have also provided valuable support and stimulus for our work here." Later, Clyde Keeler, a Harvard Medical School researcher, visited Verschuer's Institute and raved about the facility, which was adorned with numerous swastikas. Verschuer authored another book titled *Genetic Pathology*, in which he claimed that Jews disproportionally suffered from medical conditions such as diabetes, deafness, nervous disorders and tainted blood. The book received great praise in eugenics circles, and the *Eugenical News* provided a glowing review in January 1936.

Verschuer also received direct support from Charles Davenport. On December 15, 1937, Davenport asked him to prepare a summary of their work to be published in *Eugenical News*, "to keep our readers informed."[323] Davenport also urged Verschuer to join the advisory board of *Eugenical News*, which already consisted of Eugen Fischer, Ernst Rudin and Falk Ruutke. Verschuer accepted the offer, and the link between the Eugenics Record Office and eugenicists in Nazi Germany was further solidified.

All along, Verschuer had a personal assistant of whom he was quite proud. The name of that assistant was Josef Mengele, who would later be known to the world as the Angel of Death. Josef Mengele drew great inspiration from Ernst Rudin and other leading German eugenicists. Like Henry Goddard, who claimed that he could identify feebleminded individuals from a mere glance, Mengele claimed he could identify a person of Jewish ancestry by merely examining their photograph.[324] Theodor Mollison, a professor

Vol. XIX, No. 6 November-December, 1934

Eugenical News

CURRENT RECORD OF HUMAN GENETICS AND RACE HYGIENE

THE STERILIZATION LAW IN GERMANY.[1]

Dr. C. Thomalia.

In the new Germany laws are made for the benefit of posterity, regardless of the approval or disapproval of present generations. The law providing loans to married couples, for instance, taxes single persons for the benefit of the newly wed. Since repayments on these loans are automatically reduced and cancelled with the increasing number of children, this is a typical law providing for the future. The hereditary-homestead-law restores the old Germanic law of primogeniture and, therefore, helps to preserve a sound and unburdened peasant stock. The introduction of labor camps and sport organizations for the youth have made it possible to test whole classes of the growing population for biological and hereditary soundness, to observe their development and thus to select the fittest and most valuable among them. The reduction of the number of university students to only 15,000 to be admitted next year, will check the growth of an "educated" proletariat which threatened to swamp Germany in the past. Here, too, the laws of heredity are important, for young men and women who complete their university training marry much later than those who take up some other occupation.

[1] Translated from the original German by Alice Hellmer.

These are a few striking examples which prove how, in Germany, every thought and effort is directed towards improving the nation's health by applying biological laws.

In these questions, too, Germany learned from the United States. For several decades over two dozen states in America have enacted sterilization laws for the purpose of preventing hereditary degenerates like the feebleminded, the idiots, insane and those with morbid criminal tendencies to transmit their defects to their descendants. But just this law created a tremendous sensation all over the world. Scientists and laymen of various nationalities greeted it enthusiastically as a milestone in the history of mankind and as return from a hitherto wrong path. However, this law for the prevention of hereditary degenerate offspring has been criticized by some as a return to barbarism and Hunnish paganism. We Germans bear this reproach together with the enlightened, progressive United States, also with some Swiss Cantons and the Scandinavian countries; we are glad that in other countries, like Hungary, Czechoslovakia and recently England, opinions in favor of a similar law have been heard.

The reasons why nations, which are still guided by sane and calmly thinking men, have adopted such a law, are so clearly evident, that every child must understand them.

The primary and most obvious reason is that of expenditure. For a healthy

A portion of an article titled "The Sterilization Law in Germany" in vol. 19, no. 6 of the *Eugenical News* (November–December 1934). *Photo courtesy of Mark A. Torres.*

at Munich University, personally mentored Mengele. Under his guidance, Mengele earned a PhD in 1935 and soon began his medical practice in a clinic at Leipzig University. In 1937, Mengele became Verschuer's personal assistant at the Institute for the Biology of Heredity and Race Hygiene. Together, they authored opinions for eugenic courts in Germany in support of the anti-Jewish Nuremberg laws. One year later, Mengele received his medical degree. He remained in close contact with Verschuer, and both men were heralded by the Nazi party as eugenic doctors.[325]

At the onset of the Second World War, Mengele wanted to enter combat but was prevented from doing so because of a kidney condition. He continued to work closely with Verschuer and contributed to a series of book reviews and other writings on eugenics. In June 1940, Mengele joined the Waffen SS and was assigned to the Genealogical Section of the SS Race and Settlement Office in Poland.[326] Two years later, as the Nazi regime was finalizing plans for the Final Solution—to include the use of concentration camps to eliminate Jews en masse—Mengele was transferred to the SS Race and Settlement headquarters in Berlin. Mengele continued to collaborate with Verschuer, who had just accepted the director position at the Kaiser Wilhelm Institute for Anthropology, Human Heredity and Eugenics at Berlin-Dahlem. In early 1943, Verschuer had ordered the transfer of his trusted assistant to conduct work on behalf of the institute, and on May 30, 1943, Josef Mengele arrived at Auschwitz.[327]

The ability to study—or, in most cases, torture—human subjects was always the goal for German eugenicists, and above all else, human twins were their most prized subjects. This was particularly the case with identical twins, who were viewed as true clones of nature who could be tested in many conditions in search of a variety of responses. More importantly, twins were highly valued because if the hereditary genes could be manipulated, as German eugenicists believed, then breeding two perfect human beings at a time instead of one would be the best and fastest way of creating a master race.[328] Two decades earlier, Charles Davenport had recognized the hereditary value of twins. In his 1911 book *Heredity in Relation to Eugenics*, he wrote, "It is well known that twin production may be a hereditary quality."[329] In 1916, the ERO published several articles on twins in *Eugenical News*, and after circulating a four-page questionnaire for twins, it was discovered that Columbia, Missouri, had the most twins per capita in the nation, with one pair for every 447 births.[330]

The Nazi regime began its study on German *Zwillenge* (twins) in 1933, when Hitler came to power. Eugenicists conducted careful investigations, which revealed that as of 1921, there were 19,573 pairs of twins and 231 sets of triplets in Germany. In 1925, 15,741 pairs of twins and 161 sets of triplets were born. Eager to conduct further studies, the Nazi party promoted so-called twin camps in hopes of studying as many twins and triplets as possible. Over time, Verschuer and his team of eugenicists studied many groups of twins to help understand and combat hereditary diseases like tuberculosis. German neuropsychiatrist Henrich Kranz published a study on approximately 75 pairs of twins and 50 pairs of opposite-gender twins, seeking a connection with criminal behavior. These studies were praised and encouraged by the ERO in the *Eugenical News*.

Verschuer's interest in and work on twins grew exponentially. In September 1938, he secured funding for more research, and one year later, Interior Minister Wilhelm Frick issued a decree compelling all twins to register at their local health office for genetic testing.[331] Up to that point, these studies were relegated to personal observations and the taking of measurements, but Verschuer wanted more. His desire was to conduct autopsies on the bodies of twins to study their genetic material. For this, he dispatched Josef Mengele to Auschwitz in May 1943 to immediately begin directing this research.

To accommodate the large number of subjects they would study, a separate concentration camp was created for only twins, solely for the purpose of barbaric medical experimentation. As the trains arrived in Auschwitz, the terrified passengers were herded off the cars amid armed soldiers and barking dogs. The loud call of "Zwillinge! Zwillinge!" ("Twins! Twins!") rang out, and all twins were whisked away to their camp, while many of the other passengers were either worked to death or immediately sent to be gassed in the dreaded "showers." The gruesome fate that awaited twins at their camp was torture by a variety of measures that included electroshock, syringes, eye injections and more. As author Edwin Black described it, "Nazi Germany had now carried out eugenics further than any dared expect. The future of the master race that would thrive in Hitler's Thousand-Year-Reich lay in twins."[332]

At their own camp, twins were often fed well and enjoyed the personal freedom of roaming about the camp. However, and without warning, Mengele and other Nazi officers would often erupt in a violent frenzy and beat prisoners to death. Such behavior could not be more evident than in the example of twin brothers named Guido and Nino. The two boys were once given chocolates and other treats. A few days later, they were subjected to a gruesome operation where their wrists and backs were sewn together, mimicking Siamese twins, and their veins were interconnected. Their surgical wounds were left untreated and became infected. Their mother reportedly ended their agony by giving them fatal injections of morphine.[333] At times, Mengele would murder twins simultaneously to analyze them comparatively. On one occasion, he shot two boys in the back of the neck and dissected them "while they were still warm." It is believed that as many as 1,500 twins were tortured by Mengele; fewer than 200 of them survived. Josef Mengele also studied dwarfs and those afflicted with physical deformities. In one particularly barbaric case, he removed a man's stomach without administering anesthesia.

The records of all questionnaires and experiments conducted at the camp were meticulously maintained, and copies of the files were sent to Verschuer's office. At times, Mengele would send him body parts in boxes marked "War Material—Urgent."[334] On one occasion, Mengele examined a Jewish man, who was hunchbacked, and his son, who had a deformed foot. After interviewing them, Mengele had them both shot to death. The skin was boiled off their bodies, and their skeletons were sent to Verschuer at the Institute under the label "Urgent: National Defense."

The study of the eyes was always a critical part of Mengele's eugenic testing. He was keenly interested in eye color, and he conducted experiments to see if brown eyes could be changed to Nordic blue, administering dyes, sometimes by droplets and other times by direct injection. The colors of his subjects' eyes never changed, but the victims were almost always left blind. American eugenicists had always stressed the importance of studying the eyes of human twins. In March 1933, *Eugenical News* published an article titled "Heredity Eye Defects," which offered a glowing review of a new book that had a chapter on the eyes of twins. In 1936, Harry Laughlin received a request to expand the ERO's Twin Schedule to include a question on eye color.

Eugenicists also maintained a strong interest in the study of blood, particularly the blood of twins. They claimed to be searching for genetic markers of those who were likely to be carriers of defective traits. Charles Davenport spoke of the importance of this work at the second International Eugenics Congress in 1924. Three years later, a *Eugenical News* report on the opening of the Kaiser Wilhelm Institute for Anthropology, Human Heredity and Eugenics stated, "In the section on human genetics, twins and the blood groups were specially considered." On May 14, 1932, the Rockefeller Foundation in Paris dispatched a telegram to New York asking for funds to support Verschuer's research. Shortly afterward, a three-year grant of $9,000 was approved for "research on twins and effects on later generations of substances toxic for germ plasm." With funding secured for these studies,

Mengele extracted large amounts of blood from twins and other captives. He would also give blood transfusions to women using the blood of twins so that he could observe their reaction.

Hitler directed many other doctors at different concentration camps to conduct a wide range of eugenics-based research on epilepsy and other illnesses. Special breeding facilities were also designed to produce perfect babies with true Aryan ideals. Between 1937 and 1945, nearly 250,000 people were held at the Buchenwald concentration camp, located near Weimar, Germany. Throughout the war, hundreds of thousands died each week from beatings, disease, starvation and execution. Many were sent to dig tunnels into mountains until they perished from the relentless, backbreaking work. A facility infamously known as Block 46 was used for medical experiments. Prisoners here were deliberately infected with typhus to study their condition or burned with phosphorous to observe their reaction to treatment. More than a dozen were castrated, and others were victims of gland implants, injected with synthetic hormones intended to reverse homosexuality or murdered with injections of phenol.[335] Other prisoners were shot in the back of the neck through a small hole as they stood adjacent to a wall expecting to have their height measured. Pathologists dissected about 35,000 corpses for study. A Nuremberg trial judge later stated, "If there is such a thing as a crime against humanity, here we have it repeated a million times over."

Over time, the world began to learn of the Nazis' atrocities. In 1936, the Rockefeller Foundation finally became reluctant to fund any further eugenics-based programs, and nearly all funding ended when the fighting erupted in 1939. Unfortunately, Nazi eugenics programs had already benefited from the foundation's funding, and the fully developed programs continued throughout the war. The Allied invasion of 1944, coupled with the advance of the Soviet armies from the west, finally led to the defeat of the Nazi regime. Auschwitz was liberated on January 27, 1945. Josef Mengele fled to Czechoslovakia with his research documents before escaping to South America, where he lived a comfortable life until his death in 1979 from drowning. Verschuer was shielded by both German and American eugenicists and remained a respected scientist in Germany until his death in 1969 in an auto crash. He was never prosecuted for his crimes.

In the summer of 1936, Carl Schneider, dean of the University of Heidelberg Medical School, offered high praise to Harry Laughlin at the Eugenics Record Office for being "one of the most important pioneers

T R A N S L A T I O N.

Heidelberg, May 16, 1936

Dear Colleague,

The Faculty of Medicine of the University of Heidelberg intends to confer upon you the degree of Doctor of Medicine h.c. (honoris causa) on the occasion of the 550-year Jubilee (27th to 30th of June 1936). I should be grateful to you if you could inform me whether you are ready to accept the honorary doctor's degree and, if so, whether you would be able to come to Heidelberg to attend the ceremony of honorary promotion and to personally receive your diploma.

After receiving your concent we shall send you the invitation and program for the Jubilee celebration. Please regard the contents of my letter as confidential until the official announcement of honorary promotions. I should particularly appreciate your very prompt reply.

Very respectfully,

(Signed) Schneider

Dean of the Faculty of Medicine.

This page and opposite: Correspondence between Carl Schneider, dean of the Faculty of Medicine at the University of Heidelberg in Germany, and Harry Laughlin of the Eugenics Record Office in Cold Spring Harbor, New York, 1936. Schneider informed Laughlin of the school's desire to award him with an honorary degree as part of its 550th anniversary jubilee event, an award that Lauglin eagerly accepted. A short time later, Schneider helped organize the Nazi Aktion T4 program, which led to the euthanasia of thousands of mentally handicapped people throughout Germany, Austria and German-occupied Poland. *Photos courtesy of the Truman State University, Pickler Memorial Library, Special Collections and Museums.*

May 28th 1936.

Dr. Carl Schneider,
Dean of the Faculty of Medicine,
University of Heidelberg,
Heidelberg, Germany.

My dear Dr. Schneider,

I acknowledge with deep gratitude the receipt of your letter of May 16th in which you state that the Faculty of Medicine of the University of Heidelberg intends to confer upon me the degree of Doctor of Medicine h.c. on the occasion of its 550th jubilee-year 27th - 30th of June, 1936.

I stand ready to accept this very high honor. Its bestowal will give me particular gratification, coming as it will from a university deep rooted in the life history of the German people, and a university which has been both a reservoir and a fountain of learning for more than half a millenium. To me this honor will be doubly valued because it will come from a nation which for many centuries nurtured the human seed-stock which later founded my own country and thus gave basic character to our present lives and institutions.

I regret more than I can say that the shortness of time before the jubilee-date makes it impossible for me to arrange to leave my duties at Cold Spring Harbor to visit Heidelberg, to participate in the ceremony and to receive this highly honored diploma in person.

With highest regards,

Very respectfully yours,

HARRY HAMILTON LAUGHLIN

In Charge of the Eugenics Record Office,
Carnegie Institution of Washington,
Cold Spring Harbor, Long Island, N.Y.

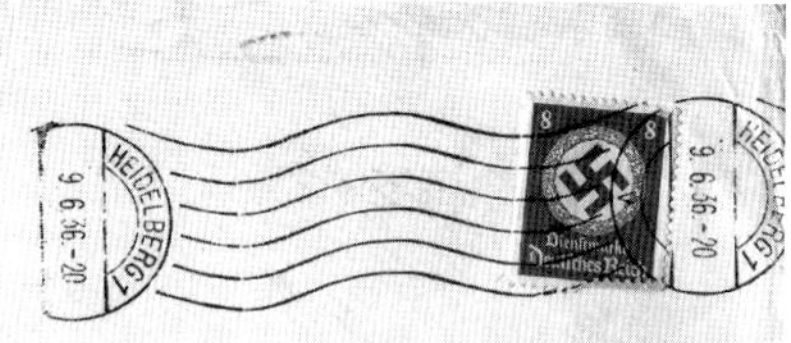

Herrn
Prof. Dr. Harry Hamilton Laughlin
Charge of the Eugenics Record Office
Carnegie Institution of Washington

Cold Spring Harbor
Long Island N.Y.

550 JAHRE
Universität Heidelberg
1386-1936

Die Universität Heidelberg

die älteste Hochschule des Deutschen Reiches, begeht in den Tagen vom 27. bis 30. Juni 1936 die Feier ihres

550jährigen Bestehens.

Ich würde es mir zur Ehre anrechnen

Herrn Prof. Harry H. Laughlin

in diesen Tagen als Gast der Universität begrüßen zu dürfen. Ihre Antwort erbitte ich möglichst bis 31. Mai 1936, damit Ihnen rechtzeitig die näheren Mitteilungen zugehen können.

Heidelberg, im Mai 1936

Der Rektor der Ruprecht-Karls-Universität

Above: Invitation from Heidelberg University in Germany to Harry Laughlin in Cold Spring Harbor, New York, with a Nazi swastika postage stamp, 1936. The invitation was to an event celebrating the school's 550th anniversary jubilee. *Photo courtesy of the Truman State University, Pickler Memorial Library, Special Collections and Museums.*

Left: Honorary doctor of medicine degree issued by the Heidelberg University School of Medicine in Germany to Harry Laughlin of the Eugenics Record Office in Cold Spring Habor, New York, for his work on eugenics. *Photo courtesy of the Truman State University, Pickler Memorial Library, Special Collections and Museums.*

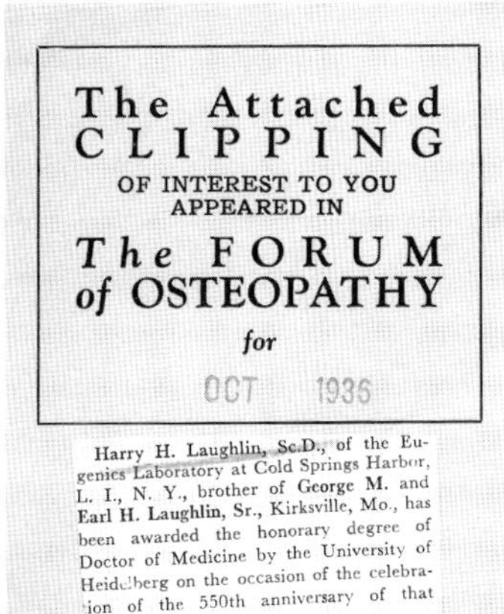

The Attached
CLIPPING
OF INTEREST TO YOU
APPEARED IN
The FORUM
of OSTEOPATHY
for
OCT 1936

Harry H. Laughlin, Sc.D., of the Eugenics Laboratory at Cold Springs Harbor, L. I., N. Y., brother of George M. and Earl H. Laughlin, Sr., Kirksville, Mo., has been awarded the honorary degree of Doctor of Medicine by the University of Heidelberg on the occasion of the celebration of the 550th anniversary of that institution.

An October 1936 article from the Forum of Osteopathy announcing the bestowal of an honorary degree from the Heidelberg University in Germany on Harry Lauglin of the Eugenics Record Office in Cold Spring Habor, New York, for his work on eugenics. *Photo courtesy of the Truman State University, Pickler Memorial Library, Special Collections and Museums.*

in the field of racial hygiene." Eager to publicly recognize Laughlin's work, Schneider wrote to Laughlin to offer him an honorary doctor of medicine degree from the University of Heidelberg at its 550th anniversary jubilee event. Although he could not personally attend the gala, Laughlin gleefully responded that he was ready to "accept this very high honor" and that "its bestowal will give me particular gratification."[336] Just three years later, Schneider served as a lead researcher to help organize the Nazi Aktion T4 program, which, at various locations throughout Germany, used carbon monoxide to lethally gas thousands of people who were deemed mentally handicapped.[337]

Named after Tiergartenstrasse 4, the address of its headquarters in Berlin, Germany, T4 was a Nazi mass euthanasia program ordered by Adolf Hitler on September 1, 1939, and spearheaded by his personal doctor, Karl Brandt, along with other doctors, in the name of eugenics and race hygiene.[338] The T4 program led to the systematic murder of approximately seventy thousand men, women and children who were disabled or otherwise deemed eugenically "unfit" German citizens in Germany.

To carry out the T4 program, several large killing centers were created throughout Germany, Austria and German-occupied Poland.[339] Many of the murdered victims were taken from poorhouses and psychiatric facilities. The methods of killing were often disguised, with many being driven in fake ambulances or buses and then herded into gas chambers disguised as showers. Once the victims were killed, their bodies were burned, and falsified death certificates were created. Later, these documents and urns with random ashes were sent to the families to disguise their loved ones' true cause of death. Throughout the Second World War, nearly three hundred thousand people were murdered as part of the Nazi mass euthanasia program.

Unlike Josef Mengele and Otmar Freiherr von Verschuer, who both escaped prosecution, Karl Brandt and twenty-two other doctors stood trial in Nuremberg, Germany, in 1946 for committing crimes against humanity.[340] To justify their acts of torture and murder, Karl Brandt defended his eugenic ideals and stated, "I took the clearly human point of view, and that is the point of view that here the life of an incurable person should be shortened."[341] As part of his defense, he cited Harry Laughlin's report and the United States Supreme Court's decision in *Buck v. Bell* to suggest that acts like his in the name of eugenics were both promoted and legitimized in the United States. The Nuremberg court ultimately rejected these arguments, and in April 1947, Karl Brandt and six others were convicted, sentenced to death and executed. On December 11, 1946, Carl Schneider hanged himself in a prison cell in Frankfurt, Germany, as he awaited trial for his actions. Later, his membership in Heidelberg's academy of science was officially rescinded.[342]

The fact that Carl Schneider bestowed an honorary degree from the esteemed Heidelberg University on Harry Laughlin in 1936 for his work on eugenics was no coincidence. It was Laughlin who drafted the *Report of the Committee to Study and to Report on the Best Practical Means of Cutting Off the Defective Germ-Plasm in the Human Population*. This report contemplated several measures for eliminating those who were deemed eugenically unfit. No. 8 on that list was mass euthanasia, and while this measure was never officially carried out in the United States, the entire report created both a mandate and a blueprint for the U.S. eugenics program that later inspired American eugenicists' counterparts in Nazi Germany. It was also Laughlin, in his capacity as superintendent of the Eugenics Record Office, who championed eugenics, race hygiene and xenophobia in the United States, Germany and throughout the world.

In 1936, Carl Schneider honored Harry Laughlin for his work on eugenics at a glorified event filled with much pageantry. A short time later, Schneider was instrumental in developing the T4 program, which led to the mass murder of nearly seventy thousand German citizens who were declared eugenically unfit—a goal originally contemplated nearly three decades earlier by American eugenicists. The methods used in the T4 program ultimately served as a precursor to the Holocaust. Perhaps more than any other example, the bestowal of this honorary degree on Harry Laughlin highlights the direct and continuous inspiration and collaboration between officials at the ERO and the murderous Nazi regime whose programs ultimately led to the death of millions of people during the Second World War.

PART III

DISCREDITED

Chapter 14

THE FALL OF THE ERO

Eugenics has come to mean an effort to foster a program of social improvement rather than an effort to discover facts. I have just observed in Germany some of the consequences of reversing the order as between program and discovery.
—L.C. Dunn, July 3, 1935

By as early as 1922, the Carnegie Institute was becoming increasingly frustrated with Harry Laughlin, who, as one of its employees, became a leading public figure associated with the xenophobic anti-immigration measures being formulated by the U.S. Congress. Nevertheless, it continued to fund the ERO's operations. In 1928, John C. Merriam, president of the Carnegie Institute, was touring Mexico with other U.S. government officials when the local press questioned him vigorously about a published story on how the United States was seeking to sharply limit Mexican immigration into the country. The story, which was amplified by Harry Laughlin, was a source of great embarrassment to the Carnegie Institute. Merriam instructed Charles Davenport to limit Laughlin's actions, but Laughlin refused to curtail his activities or public statements on immigration.[343]

In February 1929, the Carnegie Institute dispatched a committee to the ERO to review its operations. Shortly thereafter, the committee announced that the ERO's index cards, pedigree charts and family trees, some of which were decades old, amounted to little more than clutter. The committee stated that "they are of value only to the individual compiling them" and provided no real evidence on heredity. This was the first time that the

Carnegie Institute was known to have questioned the eugenics research of the ERO, and it must have come as a serious blow to Charles Davenport, who worked for so many years to develop the program. However, after several discussions, the committee decided to allow the ERO's operations to continue but demanded a closer affiliation between the Eugenics Research Association and the ERO so that more oversight could be provided.

With Hitler's rise to power in 1933, Harry Laughlin closely aligned the Eugenics Records Office, the Eugenics Research Association and *Eugenical News* with pro-Nazi propaganda. The Carnegie Institute sought to tame the content of *Eugenical News*, but the publication was the property of the ERA and Laughlin, as its secretary, resisted.[344] Two years later, the prominent geneticist C.L. Dunn began to advocate for the closure of the Eugenics Record Office. He told officials at the Carnegie Institute, "With genetics, its relations [with eugenics] have always been close, chiefly due to the feeling on the part of many geneticists that eugenical research was not always activated by purely disinterested scientific motives but was influenced by social and political considerations."[345] Later, Dunn began to ramp up his public criticism of the connection between Nazi and American eugenicists.

After the retirement of Charles Davenport in 1934, Harry Laughlin had hopes of replacing him as the director of the ERO. However, the Carnegie Institute had other plans, and in 1935, an advisory committee led by archaeologist A.V. Kidder was dispatched to, again, assess operations at the ERO and the adjacent Carnegie Station for Experimental Evolution. After a thorough evaluation of one million index cards, approximately thirty-five thousand files and other records, the committee determined that the entire operation was little more than a social propaganda machine. In a damning statement that essentially debunked all the work ever compiled by the ERO, the committee explained that the records assembled on individuals and families were

> *unsatisfactory for the scientific study of human genetics…because so large a percentage of the questions concern…traits, such as "self-respect," "holding a grudge," "loyalty," and "sense of humor," which can seldom truly be known to anyone outside an individual's close associates; and which will hardly ever be honestly recorded, even were they measurable, by an associate or by the individual concerned.*[346]

Even the material on traits that could have legitimately yielded important genetic information was so sloppily recorded by "untrained persons" that it was rendered "worthless for genetic study."[347] Ultimately, since none of the data at the ERO was ever analyzed or reviewed by the IBM data processing system, it became clear that all conclusions made by the ERO were entirely unsupported by reams of useless documents.

Stunningly, and despite these observations, the ERO was allowed to continue operating. However, amid the intensifying Nazi atrocities, the Carnegie Institute could no longer ignore the problem. In 1937, Laurence Snyder was appointed president of the Eugenics Research Association and chairman of its Committee on Human Heredity. In a dramatic departure from his predecessors, Snyder issued a lengthy report on the ERO's operations to the Carnegie Institute that signaled the beginning of the end for the program. On behalf of the committee, he wrote:

> *The recent attacks upon orthodox eugenics and indeed upon the whole present social set-up…emphasize more than ever the need for accurate facts and information on basic human genetics. These attacks, it may be stated in passing, come not from irresponsible nor untrained minds, but from some who have the authority of long and honorable scientific achievements behind them.*[348]

The report expressed deep concerns over the turmoil in Europe and lamented that in the United States, the concept and legitimacy of human genetics were being tarnished by eugenics. As a result, and to distance itself from further damage, the Carnegie Institute ordered that all references to eugenics were to be permanently replaced with "genetics." Longtime eugenicists like Davenport, Laughlin and Paul Popenoe protested and continued to openly praise eugenics. They even remained committed to the Nazi eugenic program over the years, but they must have realized that the era of eugenics was coming to an end.

On January 4, 1939, the new Carnegie president, Vannevar Bush, informed Harry Laughlin that funding to continue operations at the ERO was coming to an end.[349] When fighting erupted after the Nazis invaded Poland, Laughlin sought new sponsors to fund the ERO. However, his efforts were in vain, and on December 1, 1939, Harry Laughlin officially retired; the ERO closed permanently on the same day.

Once the ERO was officially closed, the Carnegie Institute began to dismantle the organization. It sold the building at Cold Spring Harbor but retained all other facilities.[350] In September 1947, many of the ERO's records were sent under special arrangement to the Dight Institute of the University of Minnesota and the American Philosophical Society in Philadelphia. Eugenic laws remained on the books in the United States after the ERO's closing, and although the ERO was in operation for a relatively short time (twenty-nine years), the work performed at this two-story administrative building in Cold Spring Harbor has left an egregious and enduring legacy.

Although he retired from the Carnegie Institute in 1934, Charles Davenport never wavered in his scientific support for eugenics both in the United States and in Nazi Germany. Nor did he abandon his tireless work ethic. At sixty-eight years old, he was allocated a small room at the ERO prior to its closure and some clerical assistance to continue what he called the "uninterrupted research" on mice, children and other organisms.[351] Within ten years after his retirement, Davenport wrote forty-seven papers, a new book and the fourth edition of his book *Statistical Methods*.[352]

In 1944, Davenport became very interested in the study of a whale that had beached itself off the eastern end of Long Island. He was particularly interested in retrieving the skull of the mammal for presentation at his new whaling museum at Cold Spring Harbor. He spent many nights in an open shed boiling the whale's head in a cauldron, with the bitter cold wind pressing on all sides. Colleagues recalled how he often reeked of blubber. As he continued this work, Davenport grew weaker and eventually became ill. On February 18, 1944, he died of pneumonia.[353] The veritable father of American eugenics, who dedicated his life to rooting out defective germ-plasm, did not perish from old age or sudden accident. In the end, it was infectious germs that killed Charles Benedict Davenport.[354]

In his will, Davenport left instructions for his brain to be preserved at the Wistar Institute of Anatomy and Biology in Philadelphia. The remainder of his body was laid to rest at Memorial Cemetery at St. John's Church in Laurel Hollow, New York.[355] Two years later, on March 9, 1946, his wife, Gertrude Crotty Davenport, died peacefully in a nursing home in Nyack, New York, at the age of eighty-one and was buried with her late husband. She left behind two daughters, Millia Davenport and Jane Joralemon de Tomasi.[356]

During much of his life, Harry Laughlin battled a hidden illness that he and his wife kept a secret. This illness brought on sudden seizures and once led to a near-fatal automobile crash on a road near Cold Spring Harbor. Following that accident, Laughlin never drove again. For many years, he suffered from this illness in silence before finally discovering the diagnosis while on a trip to Europe in 1920. The man who spent his entire professional career focusing on eliminating the genetic defect of epilepsy was, in fact, himself an epileptic.

Harry Laughlin spent his retirement years in Kirksville, Missouri, with his wife until his death on January 26, 1943. He was buried near his parents in a cemetery in Highland Park, in Kirkland.

Chapter 15

ENDURING LEGACY

What the island needs is not public health work, but a tidal wave or something to totally exterminate the entire population. It might then be livable. I have done my best to further the process of their extermination by killing off eight and transplanting cancer into several more.…All physicians take delight in the abuse and torture of the unfortunate subjects.
—Cornelius P. Rhoads in Puerto Rico, November 10, 1931

The Eugenics Record Office was in operation from 1910 to 1939. However, the eugenic policies and practices that were conceived and enacted from this facility far exceeded its relatively limited tenure. This chapter explores some of the eugenic practices that occurred long after the closure of the ERO, as well as more contemporary practices, all of which signify the dark and enduring impact that eugenics has made on today's society.

For much of the twentieth century, there was a large number of state-run psychiatric facilities throughout the United States. These facilities were favored by eugenicists because they gave them free access to medical records and the ability to experiment on the many patients who were housed there. As the science of eugenics began to be discredited, and other mental health conditions were discovered, the number of patients permanently housed in psychiatric hospitals across the United States continued to swell. By 1941, there were a total of 490,506 patients in state, city, county, private and veterans' hospitals across the country. By 1950, that number had grown to 577,246 patients.[357]

However, beginning in the middle of the twentieth century, the number of state-run mental health facilities began to decline sharply. This was largely due to the high operating costs of these facilities and the introduction of new medications, like Thorazine, that allowed an increasing number of patients to be treated at home instead of confined at hospitals. Between 1955 and 2000, the number of state hospital psychiatric beds declined from 339 per 100,000 of the population to 22 per 100,000 of the population, and by the year 2000, nearly all the large psychiatric facilities in New York and throughout the country had closed for good.

As the country moved away from large psychiatric centers, the percentage of mentally ill inmates in the criminal justice system rose from 16 percent in 1976 to 44 percent in 2011–12.[358] This suggests that the care provided to mentally ill patients at the large psychiatric centers was essentially absorbed by the U.S. prison system. This prompts the question: are U.S. prisons also being used for eugenic purposes?

According to a 2022 report by the *Law & Equity Journal* of the University of Minnesota Law School, between 2005 and 2013, 144 female inmates in California prisons were sterilized as a form of birth control. In most cases, the medical staff at these prisons specifically targeted and coerced pregnant inmates and repeat offenders into being sterilized without providing any counseling or other viable alternatives.[359] This practice continued until a law was passed in 2014 banning the sterilization of prison inmates throughout California.

Moreover, in 2017, a Tennessee judge signed an order that offered misdemeanor offenders thirty-day reductions to their sentences if they agreed to either undergo a vasectomy or a birth control implant. Georgia and Virginia have also offered plea deals for repeat offenders if they agreed to undergo permanent sterilization. In 2020, the news outlet Vice reported on whistleblower claims of forced sterilizations that occurred at a U.S. Immigration and Customs Enforcement (ICE) detention center in Georgia. As of November 2020, fifty-seven detainees had stated that they were pressured to undergo unnecessary gynecological surgeries. Given the perceived threats of deportation, there are likely more who have faced a similar fate but have never come forward to report it. Regrettably, this data shows that the U.S. prison system has adopted programs eerily similar to the sterilization programs designed by officials at the Eugenics Record Office and eugenicists nationwide over a century ago.

In the wake of the 1927 Supreme Court decision in *Buck v. Bell*, the number of sterilization procedures across the country grew exponentially. However, even after many states repealed their eugenic sterilization laws after World War II, coerced sterilizations without proper consent continued, including a sharp spike toward the end of the 1960s.

One reason for the sharp increase in sterilizations was the American College of Obstetricians' (ACOG) elimination of its so-called Rule of 120, an age/parity formula that doctors used to determine the need for female sterilization. According to the formula, if a woman's number of living children, multiplied by her age, equaled 120 or more, she should be sterilized. This was never legally binding, but many hospitals observed this formula. Additionally, the ACOG dismissed its recommendation for two physicians' signatures along with the rule that a psychiatric consultation be obtained before scheduling a sterilization procedure, leading to an increase in sterilizations.[360]

In the late 1960s, the U.S. government had a renewed concern over the nation's growing population. As a result of these fears, President Richard M. Nixon formed a new Commission on Population and the American Future and appointed John D. Rockefeller as chairman. President Johnson's previous War on Poverty initiative had resurrected the Malthusian belief that the world's resources were limited and would not be able to provide for the growing population. This led to political and social pressure to limit family size, and sterilization was urged by the newly formed Office of Economic Opportunity, an agency that sought federal funds to provide education and training to the poor, along with the introduction of more modern forms of contraception.

As a result of this initiative, the Family Planning Act of 1970 was overwhelmingly passed by Congress. The combination of this new legislation and changes in medical practice had a dire impact on minority women. During the 1970s, the U.S. Department of Health, Education and Welfare (HEW) funded 90 percent of annual sterilization costs for poor people. Continuing the legacy created long before by eugenicists, physicians and social workers were empowered to exercise their authority over the reproductive rights of poor and minority families. As a result, sterilization for women "increased by 350 percent between 1970 and 1975, and approximately one million American women were sterilized each year."[361]

CALIFORNIA HAD ALWAYS LED the nation in the number of eugenic sterilization procedures performed. From 1909, when its first eugenic sterilization statute was enacted, to 1921, there were approximately 2,558 sterilizations in the state.[362] By the end of World War II, the total number of sterilizations performed in the state had reached nearly 15,000, and that number grew to over 20,000 by 1983.

During the 1960s and '70s, Los Angeles County–USC Medical Center was one of the largest public hospitals and medical training centers in the United States. The hospital provided care for low-income men, women and children, many of whom were of Mexican descent. However, it was later discovered that between 1968 and 1974, hundreds of female patients were sterilized without their knowledge and against their will.[363] Such procedures were directly eugenic in nature.

The heart-wrenching 2015 documentary *No Más Bebés* covers this untold history. The film describes how patients who were in labor were coerced to sign a consent form that granted doctors at the hospital permission to perform a tubal ligation.[364] In most cases, through their inability to read or understand the consent forms, which were always in English, coupled with the physical pain they were in, patients believed they had had no alternative but to sign the forms—and in so doing, they unknowingly gave away their ability to ever have children again.

Bernard Rosenfield was a doctor at the hospital who later became a whistleblower about the widespread practice. In the film, he explains that doctors were instructed to actively urge female patients to have the sterilization procedure, regardless of their age. He also recounts how one doctor stated that he "picked up a whole new prejudice" about Mexican women and others routinely ridiculed their patients for being promiscuous and having too many children. Other witnesses in the film add that little explanation of the procedure was ever provided to patients, and many of them mistakenly believed that the procedure was either reversible or merely a "cleaning." One woman did not even know she had been sterilized until four years later, and another who was writhing in pain was told that in order to receive an injection to alleviate the pain, she had to sign the consent form.

The film also tells the story of Dolores Madrigal, a woman who was sterilized without providing her consent. Along with never being able to have children again, news of the procedure adversely affected her marriage, as her husband accused her of having the procedure purposefully so that she could be promiscuous. As a result, he began drinking heavily and began to abuse her mentally and physically. She was so distraught that she even attempted suicide. In the film, social workers explain that many of these women exhibited symptoms similar to those of post-traumatic stress disorder after the procedure and were shunned by their families and communities.

In 1978, a young attorney named Antonia Hernández took on the Madrigal case and filed a class action civil lawsuit on behalf of ten plaintiffs against the hospital, its administrators and doctors.[365] The complaint alleged that the plaintiffs' constitutional rights were violated when the hospital subjected them to sterilization procedures without their voluntary and informed consent as part of a concerted plan to prevent women in low socioeconomic groups from having large families. The case drew large media attention during a time when women's reproductive rights was being debated across the country.

On June 30, 1978, U.S. District Judge Jesse W. Curtis dismissed the case by ruling that the doctors had the best interests of the patients in mind when they performed the sterilization procedures. Although the plaintiffs lost the case, some positive changes followed, which included the hospital's agreement to provide bilingual consent forms and to Spanish-speaking counselors on staff to communicate with patients.

On March 11, 2003, California governor Gray Davis offered a public apology for the state's participation in eugenics. He noted that California was part of a small group of states that have expressed similar regrets. In a prepared statement, he said, "To the victims and their families of this past injustice, the people of California are deeply sorry for the suffering you endured over the years. Our hearts are heavy for the pain caused by eugenics. It was a sad and regrettable chapter…one that must never be repeated."[366]

In 2018, the Los Angeles County Board of Supervisors also issued a public apology for the sterilization of more than two hundred women at LA–USC Medical Center between 1968 and 1974. In a further attempt to reconcile with its dark past, California became the third state in the nation to approve a reparations program for victims (or, if deceased, their beneficiaries) of forced eugenic sterilizations in its hospitals and prisons.[367] Passed on July 10, 2021, the law set aside $7.5 million to pay reparations to the victims and approximately $1 million for plaques to honor them. The program specified two groups of victims: those sterilized during the 1930s (the peak period of sterilizations in the state) and a smaller group who were sterilized in prisons within the past ten years.

The process of paying reparations got off to a slow start, as the search for living victims from the first group proved to be difficult. Within the first year of the law's passage, just 51 people were paid out of 310 applicants. As of 2023, there were 153 applications being processed, while 103 applications were denied for being incomplete or containing erroneous information. Advocates placed posters and fact sheets in various nursing homes and libraries across the state, hoping to locate further victims.

The State of California passed its first eugenic sterilization law in 1909. The law targeted captive populations that eugenicists wanted to prevent from procreating through a mass sterilization program. More than one hundred years later, the state once again was targeting similarly captive populations in a manner eerily reminiscent of the heyday of the eugenics era. Thus, while the state issued a formal public apology and passed a reparations law to compensate victims, it is apparent that the legacy of eugenics is enduring.

THE STATE OF NORTH Carolina passed its own eugenic sterilization law in 1929. While most states ceased to perform or performed fewer eugenic sterilizations after World War II, the rate of sterilizations in North Carolina greatly accelerated after the war. By 1950, there had been nearly 2,500 sterilizations performed, and by 1963, that number had grown to over 6,000. As of 1978, the total number of men, women and children who had been sterilized in the state is estimated to have been 7,800, with almost all of the procedures involuntary.[368]

For many years, little was known about North Carolina's eugenical sterilization program—until it was stumbled upon by historian Johanna Schoen, who was conducting research on birth control and reproductive health in the state.[369] Having discovered the breadth of the program, Schoen shared her findings with reporters from the *Winston-Salem Journal*, who covered the story in a series of articles in 2002. In the wake of the reporting, Governor Mike Easley issued the following statement on December 13, 2002: "I deeply apologize to the victims and their families for this past injustice.... This is a sad and regrettable chapter in the state's history, and it must be one that is never repeated again."

In 2016, an award-winning journalist named Dawn Sinclair Shapiro produced a riveting documentary on this topic titled *The State of Eugenics*.[370] The film exposes how, in true eugenic fashion, the victims who were targeted for these procedures were the poor and the powerless. After a brief evaluation by the state eugenics board, patients were recommended for sterilization or asexualization (a term used to describe the complete removal of reproductive organs). The records also show that doctors often lied to patients about the nature of these procedures. Many were told that the sterilization procedure was temporary before undergoing the irreversible operation. One person interviewed in *The State of Eugenics* asserted that the eugenics board was "treating human beings like some other cog in the business of quality control."[371]

The records show that many state officials in North Carolina pressured individuals to agree to be sterilized or risk losing state benefits, including welfare. They also indicate a post–World War II shift of the state eugenics board from being a mere review board to a strong and persistent advocate for sterilization. In one instance, in 1947, the board sent a series of weekly letters advocating for the sterilization of fourteen-year-old Willis Lynch, who was confined to the Caswell Training School in Kinston, North Carolina, a home for orphans, the poor and those with mental disabilities. After repeated pressure from state officials, the board members of the facility approved the operation, which took place on May 12, 1948. Ultimately, more than six hundred children who occupied this facility were sterilized.

Records also show the relentless efforts of Wallace Kuralt, a social services director for the state, who often toured poverty-stricken areas with other elected officials to make the strong economic argument for sterilization. He once noted, "These families move closer to poverty with each additional child." In Mecklenburg County alone, 485 men, women and children were sterilized, a rate three times greater than in any other county.

In 1965, Nial Ruth Cox was sterilized in North Carolina. She was told that her tubes would be tied temporarily and that in exchange for agreeing to the procedure, she could continue to receive welfare assistance. The American Civil Liberties Union filed a lawsuit on her behalf, and although the litigation ultimately led to the permanent shutdown of the state's eugenics program, no compensation was offered to any victims.

In 2009, Democratic state representative Larry Womble endorsed a bill to establish compensation for the victims of eugenic sterilization in North Carolina. Unfortunately, the bill was subjected to legislative wrangling and delays before being rejected. Two years later, the Republican Speaker of the House, Thom Tillis, joined Womble to sponsor a bill. Reporter and advocate John Railey began to publish weekly editorials in support of the bill and promoting victim compensation. He wrote, "Our state is in need of redemption. We were the worst of the worst, and we can be the best of the best." Shortly thereafter, a panel on compensation for victims was established, and national media attention began to grow.

In 2012, the House passed a bill that would provide $50,000 per victim to an unlimited number of victims, but the Senate rejected the bill, as it appeared to be too open-ended. Tillis and Womble continued to work diligently to garner more support. Tillis proposed that in lieu of a bill, compensation for victims would be allocated in the state's $20 billion budget. On July 24, 2013, the budget was passed, and $10 million was set aside to be split between the living victims of sterilization. After years of waiting for justice, many of the victims received $20,000 each. In November 2015, the state made the final payments to 220 men and woman who were sterilized.

The State of Virginia has its own dark history of eugenic sterilization. On March 20, 1924, the Virginia Eugenical Sterilization Act was signed into law. It declared that "heredity plays an important part in the transmission of insanity, idiocy, imbecility, epilepsy, and crime." Although there were very few such procedures performed at the time, the number of sterilizations in Virginia grew steadily after the 1927 *Buck v. Bell* Supreme Court decision and increased in each subsequent decade. By 1938, nearly 3,000 sterilizations had been performed in the state, and by 1979, there had been nearly 7,500 procedures performed in Virginia.[372]

In an attempt to make amends for this history, Virginia passed a victim compensation bill in 2015 for those who were sterilized in the state. The new bill, which was similar to the one passed in North Carolina, set aside a total of $400,000, with approximately $20,000 for each victim or their estate.[373] Unfortunately, only eleven of the thousands of victims were still living at the time of the bill's passage.

In August 1970, Norma Jean Serena, a Native American of Creek and Shawnee ancestry, was living in Apollo, Pennsylvania. She was pregnant at the time and living with her four-year-old son, Gary, and three-year-old daughter, Lisa. Agents of the local child welfare service came to her home and abruptly took custody of the children after determining that they were malnourished. Later that month, Norma Jean gave birth to her son Shawn. Immediately after she gave birth, doctors sterilized her and placed Shawn into foster care. Norma Jean signed a consent form the day after the surgery took place. She would not see her children again for three years, until a jury determined that the social services workers had removed her children under false pretenses. In November 1970, a twenty-six-year-old woman of Native American ancestry entered a physician's office in Los Angeles to request a "womb transplant." On examining her, the doctor determined that she had previously been sterilized by means of a hysterectomy. The woman, who badly wanted to have a family, was distraught to learn that she could never bear children. These are but two of the horrible stories concerning the sterilization of Native American woman.

In 1977, Marie Sanchez, a chief tribal judge of the Northern Cheyenne Reservation, delivered a powerful message to the United Nations Convention on Indigenous Rights detailing how Native American women were systematically being targeted in a modern form of genocide.[374] After the passage of the Family Planning Services and Population Act of 1970, between 1970 and 1976, it is estimated that between 25 percent and 50 percent of Native American woman of childbearing age were subjected to eugenic sterilization. Judge Sanchez explained how one doctor chastised several women for having too many children, while others were told that the sterilization procedure was reversible.

Dr. Connie Pinkerton-Uri, a physician of Choctaw and Cherokee descent, inspected hospital records and learned that in the month of July 1974 alone, there were forty-eight sterilizations performed, and several hundred had been performed in the previous two years. Another study conducted on the Navajo Reservation between 1972 to 1978 revealed that the percentage of sterilization had more than doubled from 15.1 percent in 1972 to 30.7 percent in 1978.[375] In Vermont, the Abenaki Native people were also subjected to a harsh eugenic sterilization program.

Growing concerns over sterilization in Indian country compelled Senator James Abourezk, chairman of the Senate Subcommittee of Indian Affairs in South Dakota, to order the General Accounting Office (GAO) to conduct a thorough investigation. On November 6, 1976, the GAO released a report of the records at four of the twelve Indian Health Services (IHS) program centers: Aberdeen, Albuquerque, Oklahoma City and Phoenix. According to its report, between the years of 1973 and 1976, 3,406 Native American women were sterilized. Of this number, 3,001 were women of childbearing age (between fifteen and forty-four), and 1,024 of the sterilizations were performed at IHS contract facilities. Although the records of only four of the twelve IHS centers were examined over a forty-six-month period, these findings were staggering. Senator Abourezk noted that the 3,406 sterilizations among this much smaller population were comparable to sterilizing 452,000 non-Indian women.[376]

The report also showed that thirty-six of the women sterilized were under the age of twenty or were adjudicated to be mentally incompetent. Moreover, the consent forms obtained prior to the procedures did not meet the federal regulations for "informed consent" and were often given without the presence of an interpreter. In some cases, the required seventy-two-hour waiting period between signing the form and performing the surgery was not observed.[377]

The sterilization program of Native American women mirrors the eugenic sterilization programs of the early twentieth century. Each group represents a minority of the overall population; they each have been confined as a captive population to be studied and practiced on; and they each have been targeted for the purposes of limiting or eliminating their overall population based on eugenic ideals. Author Sally Torpy summarized the tragic sterilization of Native American woman as follows:

> *Tribal dependence on the federal government through the Indian Health Service (IHS), the Department of Health, Education, and Welfare (HEW), and the Bureau of Indian Affairs (BIA) robbed them of their children and jeopardized their future as sovereign nations.*[378]

Eugenicists thrived on testing captive populations in places like prisons, psychiatric institutions and Native American reservations. In these settings, the records were easily accessible, the subjects were controllable and the authoritative figures were typically more than eager to assist. However, one of the greatest concentrated efforts to implement a mass eugenic program may have occurred on the island of Puerto Rico, where eugenicists undertook efforts to inflict mass sterilization on an entire population.

Approximately 1,600 miles off the coast of the mainland United States, Puerto Rico was claimed as a U.S. territory in 1898 through the Treaty of Paris, which ended the Spanish-American War.[379] The 100-mile-long, 35-mile-wide Caribbean island is blessed with scenic beauty, natural resources and a proud population with a strong cultural and national heritage.

The United States utilizes the island as a hub for the shipment of goods and as a strategic U.S. naval station. American pharmaceutical companies have also made a manufacturing home in Puerto Rico. As of 2008, Puerto Rico was the largest shipper of pharmaceuticals in the world, accounting for 25 percent of total shipments. Sixteen of the twenty most popular drugs were produced on the island. For many years, Pfizer operated a factory in the Puerto Rican town of Barceloneta that produced large quantities of Viagra to treat erectile dysfunction, with annual sales exceeding $1 billion and drawing a profit of nearly 90 percent per pill.[380] Barceloneta was widely known as Viagra City (Cuidad Viagra) for many years. It embodies the dominance that the U.S. pharmaceutical industry once held over the island.

Over time, the United States grew concerned about overpopulation on Puerto Rico. As a result, government officials worked in tandem with eugenicists to develop a cruel mass sterilization program on the island. In true eugenic fashion, the program was fueled by racism and intolerance and was designed to decrease an entire population of free people.[381]

Mass sterilizations are believed to have begun in the 1930s after Menendez Ramos, the acting governor of Puerto Rico, implemented a sterilization program for women on the island.[382] He claimed that the program was needed to battle rampant poverty and economic strife. At one point, more than one third of the female population was sterilized and/or used to test birth control products without their informed consent. In the town of Barceloneta, more than twenty thousand women are believed to have been sterilized in a program known locally as the Operation (La Operación). The sinister program was so successful that in 1938, Public Laws 116 and 136 were passed, which legalized sterilization for all of Puerto Rico, even for "non-medical" reasons. These laws encouraged the island's health commissioner to "teach and practice eugenic principles."[383]

By the 1950s, Puerto Rico was facing a struggling economy. As a result, the U.S. government enacted Operation Bootstrap, an initiative to transport Puerto Ricans from the island to the mainland United States as contract laborers to perform agricultural work. At the insistence of eugenicists, who argued that Puerto Rican women engaged in "reckless breeding," plans for more sterilization were included in this initiative.[384] Official surveys revealed staggering data from this mass sterilization program. According to the study, 7 percent of the island's women had been sterilized by 1948.[385] By 1954, the number of sterilized women had more than doubled to 16 percent, and by 1965, one third of all women aged twenty to forty-nine had been sterilized. Women younger than twenty or older than forty were rarely sterilized. This is consistent with eugenic ideals, which placed a greater focus on sterilizing women during peak childbearing years. In no other industrialized or developing country has the rate of sterilization been so high. Since these surveys were written in English, and the term *sterilization* was not properly defined or understood by those who were surveyed, it's likely that the total number of sterilizations was higher. By the 1970s, it's estimated that an additional twenty-nine thousand sterilizations had been performed, more than 90 percent of them on women.

Prior to 1960, sterilizations in Puerto Rico were not governed by the U.S. Congress. Instead, such decisions were generally left to individual government workers and physicians, who were mostly White men. One of these was a physician named Cornelius P. Rhoads (1898–1959). Rhoads graduated from Harvard Medical School in 1924. He taught pathology at Harvard before joining the Rockefeller Anemia Commission, which set up a research laboratory at the San Juan Presbyterian Hospital. In 1931, Rhoads arrived on the island to work in the laboratory. If there is any doubt about the racist animus against Puerto Ricans throughout the history of its mass sterilization programs, one should look no further than the following letter written by Dr. Rhoads to a colleague on November 10, 1931, which read in part:

> *I can get a fine job here and I am tempted to take it. It would be ideal except for the Porto Ricans* [sic]*—they are beyond doubt the dirtiest, laziest, most degenerate and thievish race of men ever to inhabit this sphere. It makes you sick to inhabit the same island with them. They are even lower than the Italians. What the island needs is not public health work, but a tidal wave or something to totally exterminate the entire population. It might then be livable. I have done my best to further the process of their extermination by killing off eight and transplanting cancer into several more….All physicians take delight in the abuse and torture of the unfortunate subjects.*

As evidenced by his own words, Rhoads harbored a deep contempt for the inhabitants of Puerto Rico. His letter was published in local newspapers and drew immediate public outrage. Copies were also sent to the governor, the League of Nations, the Pan-American Union, the American Civil Liberties Union and foreign embassies around the world. They were offered as evidence of systemic and lethal U.S. racism toward Puerto Ricans. Rhoads attempted to downplay the contents of the letter by calling it satire and a "fantastic and playful composition, written entirely for my own diversion," but there is no doubt that these were his own written words and thoughts.

Despite apparently confessing to murder and medical atrocities, Rhoads suffered no adverse effects from his notorious letter or his bigotry. He continued to work and supervise hundreds of chemists, technicians, librarians and lab assistants over a long career. He later directed the cancer program at Sloan-Kettering Institute in New York. On June 27, 1949, he was featured on the cover of *TIME* magazine, where he was hailed as a "cancer fighter." Despite the accolades he received, the stated words and apparent confessions of medical atrocities committed by Dr. Rhoads and others serve as a stark reminder of the virulently racist views held by many doctors and eugenicists alike. As a direct result of such views, thousands of people in Puerto Rico were subjected to irreversible sterilization procedures for decades in a sinister attempt to control the island's entire population.

In 2002, the Netherlands became the first country to allow doctors to euthanize patients at their request if they met certain requirements, including having an incurable illness causing "unbearable" physical or mental suffering.[386] Between 2012 and 2021, an estimated sixty thousand people were euthanized under the law at their own request, according to the Dutch government's euthanasia review committee. To show that the rules were being applied properly, the committee released documents related to more than nine hundred of those people, most of whom were older and had conditions including cancer, Parkinson's and ALS. Despite the intended safeguards and strict limitations, the notion of facilitating the elimination of patients at any cost is inextricably linked to eugenic ideals.

According to a report released by the National Women's Law Center on January 24, 2022, there are currently thirty-one U.S. states that still allow for forced sterilization of people with disabilities.[387] These states include California, Florida, New York and Washington, D.C.[388] Under these laws, a judge has the authority to decide if a person can be sterilized if it is determined that it would be in the best interest of the person.

Finally, in 2023, the U.S. Food and Drug Administration began to review newly developed technology that can produce artificial wombs for human beings.[389] These are designed to treat infants who have been born prematurely. Some companies have already developed specialized transparent growth pods, which can be used as artificial wombs for would-be parents.[390] These pods are designed to simulate the conditions of an actual womb, and the growth of the fetus can be monitored on a computer or cell phone application. While this technology may be used for positive results, the dark history of eugenics should remind us of the dangers to humanity that can follow if it is used only for the strong, healthy and wealthy while disregarding the weak, sick and poor.

Chapter 16

A RECKONING

Our support of eugenics made us complicit in driving decades of brutal and unconscionable actions by governments in the United States and around the world.
—Eric Isaacs, president of the Carnegie Institution for Science, 2020

The rise of eugenics was not a random phenomenon. Eugenics was presented as a cutting-edge science driven by utopian ideals for the betterment of humanity. It was buoyed by a continuous flow of financial support from wealthy and progressive-minded donors and fully embraced by the leading thinkers of the time before settling into the very fabric of the United States and societies throughout the world. Ultimately, eugenics was discredited as a science and exposed as nothing more than a social philosophy used as a slogan for intolerance, racism, bigotry and classism. It was essentially a means for the wealthy to assert their dominance over the poor, which has been an unfortunate and recurring theme throughout all of human history.

It took many years for the scientific and corporate communities to accept responsibility for their part in eugenics. In a published statement issued during the summer of 2020, Eric Isaacs, president of the Carnegie Institution for Science, issued a formal apology for the group's support for eugenics.[391] The statement, which acknowledged the institution's relationship with Charles Davenport and the Eugenics Record Office, read in part:

There is no excuse, then or now, for our institution's previous willingness to empower researchers who sought to pervert scientific inquiry to justify their own racist and ableist prejudices. Our support of eugenics made us complicit in driving decades of brutal and unconscionable actions by governments in the United States and around the world. As the President of the Carnegie Institution for Science, I want to express my sincere and profound apologies for this organization's past involvement in these horrific pseudoscientific activities.

Some of the states who participated in brutal eugenic practices have also sought to make amends by passing legislation to compensate the victims of these practices within their states.

On January 24, 2023, the American Society of Human Genetics issued the following public statement:

The American Society of Human Genetics seeks to reckon with, and sincerely apologizes for, its involvement in and silence of the misuse of human genetics research to justify and contribute to injustice in all forms.[392]

This apology came in the wake of a twenty-seven-page report in which the group scrutinized its own past and announced recommendations to repair the damage caused for so many years.[393] This acknowledgement was issued two decades after the Human Genome Project revealed that human beings share 99.9 percent of their genetic material, thus thoroughly debunking eugenics and confirming that "race" is purely a social construct.[394] It also came amid a rise in hate crimes and mass killings inspired by White supremacy and false claims about the demographic makeup of the United States.

Accepting responsibility for one's own actions is always a positive step in the right direction. Unfortunately, these apologies came far too late and can never fully atone for the harsh treatment and cruel denial of progeny afflicted on thousands of people for so many years. Thus, a true reconciliation for the brutal legacy of eugenics can never be realized. Moreover, although we are a century removed from the onset of the American eugenics movement that once dominated our scientific, political and cultural objectivity, some practices that continue today are dangerously reminiscent of that dark period in our history. This should give us all a reason for grave concern.

The desire for the overall improvement of humanity will never cease. However, such aspirations always seem to blind us to what we inherently know or certainly should have always known: that the so-called perfect standards of our species come in many forms. While beauty, intelligence and strength are easily witnessed, there are a vast number of other less apparent but no less important qualities that show our true greatness. These are often seen in diverse families with blended cultures. They shine brightly in the face of one who overcomes unbearable odds or in the stature of one who has been reformed. They radiate in our unbounded depths of love, mercy and tolerance. And the compassion of a caregiver, particularly of one afflicted with disease, disability or old age, is unrivaled.

Ultimately, the greatness of humanity cannot be engineered in one single form, shape, color or creed, and it certainly cannot be manufactured in a laboratory or biased think tank. To the contrary, our greatness is apparent when we factor in—not seek to eliminate—our uniqueness, differences, imperfections and aspirations that are free of corruption or nefarious motives. These are the ideals that make human beings special, both individually and collectively. The American novelist Toni Morrison once said, "No more apologies for a bleeding heart when the opposite is no heart at all. Danger of losing our humanity must be met with more humanity." At its very core, eugenics was devoid of diversity, compassion, tolerance and, ultimately humanity, which is precisely why it was completely flawed and dangerous.

In the nearly three decades of its operation, the Eugenics Record Office served as the ultimate vessel to fortify and amplify the pseudoscience called eugenics and transformed it into a global phenomenon. Everything that emanated from this facility served to dominate the poor, the weak and the sick, who were deemed the defectives of society, and subject them to mass levels of institutionalization, sterilization, immigration restrictions and even euthanasia. Later, in the hands of the Nazi regime, eugenics was openly used as a scientific excuse to torture and murder a multitude of innocent human beings.

The Eugenics Record Office and those who directly operated, controlled and funded it are fully deserving of the blame for the entire eugenics movement and the dire atrocities committed under the banner of this false science. While we must continue to honor the seemingly countless victims, we must also provide public discourse and educational programs on the subject, for if we fail to do so, we may be in danger of repeating this dark history.

NOTES

Introduction

1. HLS Nuremberg Trials Project, "*U.S.A. v. Karl Brandt et al.*"
2. Cohen, *Imbeciles*, 239.
3. Singleton, "Science of Eugenics," 122.
4. Cohen, *Imbeciles*, 43.
5. Ferrari, *Eugenics Crusade*.
6. Cohen, *Imbeciles*, 269.

Chapter 1

7. Singleton, "Science of Eugenics," 122.
8. Davenport, *Heredity*, 1.
9. Ibid., 7.
10. Cohen, *Imbeciles*, 121.
11. Black, *War Against the Weak*, 10.
12. Ibid., 11.
13. Ibid., 11–12.
14. Ibid., 12.
15. Kevles, *In the Name of Eugenics*, 41–42.
16. Black, *War Against the Weak*, 26.
17. Kevles, *In the Name of Eugenics*, 5.
18. Black, *War Against the Weak*, 14.
19. Ibid., 15.
20. Ibid., 16.

Chapter 2

21. Cold Spring Harbor Laboratory, "Charles B. Davenport."
22. Ibid.
23. Kevles, *In the Name of Eugenics*, 45.
24. Dennert, "Henry Herbert Goddard."
25. Gur-Arie, "Harry Hamilton Laughlin."
26. University of Missouri, "Harry Laughlin: Workhorse."
27. *New York Times*, "Mrs. E.H. Harriman Dies."
28. Kevles, *In the Name of Eugenics*, 54.
29. Ibid., 54.
30. Ibid., 55.

Chapter 3

31. Loving-Long-Island.com, "Cold Spring Harbor."
32. *Long-Island (NY) Star*, May 26, 1819.
33. Ibid.
34. Social Networks and Archival Context, "Brooklyn Institute of Arts and Sciences."
35. *Brooklyn (NY) Daily Eagle*, July 8, 1890.
36. *Long-Islander* (Huntington, NY), April 23, 1898.
37. Ibid.
38. *Brooklyn Daily Eagle*, December 25, 1910.
39. Kevles, *In the Name of Eugenics*, 51.
40. Ibid.
41. Ibid.
42. Ibid.
43. American Philosophical Institute "Official Record of the Gift." Today the property is owned and operated by the Cold Spring Harbor Laboratory.
44. Black, *War Against the Weak*, 105.
45. Cold Spring Harbor Laboratory, "Eugenics Record Office."
46. Black, *War Against the Weak*, 105.
47. *Long-Islander* (Huntington, NY), April 4, 1913.
48. Ibid., June 9, 1911.
49. *Brooklyn Daily Eagle*, September 29, 1921.
50. *County Review*, August 25, 1922.
51. *East Hampton (NY) Star*, June 7, 1912.
52. *Northport (NY) Observer*, November 1, 1929.
53. *Long-Islander* (Huntington, NY), November 8, 1929.

54. *Northport (NY) Observer*, January 10, 1930.
55. American Philosophical Society, B: D27-2, Box 126.
56. Riddle, *Biographical Memoir*, 91.
57. American Philosophical Society, B: D27, Folder 5, Box 88.

Chapter 4

58. Kevles, *In the Name of Eugenics*, 64.
59. Okrent, *Guarded Gate*, 241–42.
60. Ibid.
61. "Eugenics in the Colleges," 186.
62. Bresnan, "Eugenics at William & Mary."
63. Paul K. Longmore Institute on Disability, "Eugenics on Campus."
64. Okrent, *Guarded Gate*, 242.
65. Ibid., 249.
66. Cohen, "Harvard's Eugenic Era."
67. Ibid.
68. Kevles, *In the Name of Eugenics*, 71.
69. Pedigrees of poor, rural families—with fictitious names like the Nams and the Ishmaelites—were utilized to show supposed genetic deterioration over time. Defining this as "cacogenics," American eugenicists dismissed any environmental contribution to their plight, such as poor housing, nutrition or schools, and instead placed the blame solely and directly on bad genes. See Cold Spring Harbor Laboratory, "Nam Family Members."
70. Cold Spring Harbor Laboratory, "Nam Family Members."
71. Cold Spring Harbor, Box 14: Series 4 ERA Subseries ERO Memoirs, 1912.
72. Black, *War Against the Weak*, 315.
73. Harry Laughlin Papers at the University of Missouri at Kirksville.
74. Lombardo, "Medicine, Eugenics, and the Supreme Court."
75. American Philosophical Society, "Letter to Charles B. Davenport."
76. *Glen Falls (NY) Times*, August 9, 1933.
77. American Philosophical Society, AES, 57506: Am3, Sermon #56.
78. Black, *War Against the Weak*, 88.
79. Nursing Clio, "Black Politics of Eugenics."
80. Archbold, "7 Beloved Famous People."
81. National Park Service, "Panama-Pacific International Exhibition."
82. Stern, *Eugenic Nation*, 48–49.
83. Hoff, "International Eugenics Congresses."

84. Black, *War Against the Weak*, 237.
85. Ibid.
86. Kevles, *In the Name of Eugenics*, 61.
87. American Philosophical Society, AES, Am3,575.06, "Proposed Itineraries."
88. Kevles, *In the Name of Eugenics*, 62.
89. Ibid.
90. Ibid.
91. Black, *War Against the Weak*, 125.
92. Ibid.
93. Ibid., 126.
94. Ibid., 127.
95. Sanger, *Pivot of Civilization*, 48.
96. Ibid., 51.
97. Black, *War Against the Weak*, 132.
98. Ibid., 133.
99. Black, *War Against the Weak*, 133.
100. Ibid., 134.
101. Ibid., 135.
102. Sanger, *Pivot of Civilization*, 49.

Chapter 5

103. Black, *War Against the Weak*, 186.
104. American Philosophical Society, "Letter to Alexander Graham Bell."
105. Wikipedia, "Madison Grant."
106. Okrent, *Guarded Gate*, 204.
107. Ibid., 208.
108. Black, *War Against the Weak*, 90.
109. Ibid., 251–52.
110. Ibid., 259.
111. Okrent, *Guarded Gate*, 209.
112. Black, *War Against the Weak*, 90.
113. American Philosophical Society, "Minutes of Meeting."
114. Black, *War Against the Weak*, 159.
115. Cohen, *Imbeciles*, 131.
116. Black, *War Against the Weak*, 189.
117. Ibid., 190.
118. Memo requested by Merriam, Harry Laughlin Papers at the University of Missouri at Kirksville, c-4-5-6.

119. Using strikingly derogatory language, the full name of the report was *Analysis of the Metal and Dross in America's Modern Melting Pot: The Determination of the Rate of Occurrence of the Several Definite Types of Social Inadequacy in Each of the Several Present Immigration and Native Population Groups in the United States*. See Laughlin, *America's Modern Melting Pot.*
120. Black, *War Against the Weak*, 190.
121. Ibid., 192.
122. Ibid., 199.
123. Ibid.

Part II

124. Singleton, "Science of Eugenics," 122.

Chapter 6

125. Black, *War Against the Weak*, 89.
126. Ibid., 57.
127. Ibid., 58.
128. Laughlin, "Preliminary Report," 17.
129. Black, *War Against the Weak*, 58.
130. Davenport, *Heredity*, 9.
131. Ibid., 18–19.
132. Laughlin, "Preliminary Report," 46.
133. Ibid., 45–56.
134. Black, *War Against the Weak*, 90.
135. Ibid.
136. Kevles, *In the Name of Eugenics*, 56.
137. Black, *War Against the Weak*, 98.
138. Ibid., 105.

Chapter 7

139. Brittanica, "Scientific Method."
140. Kevles, *In the Name of Eugenics*, 46.
141. Black, *War Against the Weak*, 52.
142. Davenport, *Trait Book.*
143. Ibid., 3.
144. Kevles, *In the Name of Eugenics*, 55.

145. Ibid., 56.
146. Cold Spring Harbor, Field Workers of the Eugenics Record Office.
147. Black, *War Against the Weak*, 52–53.
148. Cold Spring Harbor, Box 9: Folder 9.18.
149. Cold Spring Harbor, Box 9: Folder 9.25.
150. Cold Spring Harbor, Box 9: Folder 9.28.
151. Cold Spring Harbor, Box 9: Folder 9.29.
152. Cold Spring Harbor, Box 9: Folder 9.1.
153. Cold Spring Harbor, Box 9: Folder 9.32.
154. *Corrector* (Sag Harbor, NY), June 6, 1874.
155. *East Hampton (NY) Star*, June 27, 1913.
156. American Philosophical Society, ERO, MSC77, SerX, Box 3: Harry M. Laughlin.
157. Davenport and Steggerda, "Shinnecock Indians."
158. Strong, "Legacy of Long Island's First Peoples."
159. Strong, "Miss-measuring the Unkechaug," 20.
160. Ibid., 24.
161. Ibid., 21.
162. American Philosophical Society, Box Series I-63, Folder 1.
163. American Philosophical Society, ERO, Box I-63, Folder 1.
164. Ibid., 53.
165. Cold Spring Harbor, Box 14: Series 4 ERA Subseries ERO Memoirs, 1912.
166. Vassar Encyclopedia, "Dr. Katharine Bement Davis."
167. Lombardo, *Three Generations, No Imbeciles*, 240.
168. Cohen, "Historical Facts."
169. *Brooklyn Daily Eagle*, March 29, 1914.
170. Levine, "Real History of Letchworth Village."
171. American Philosophical Society, Box 63: B: D27-2.
172. Lombardo, "Tracking Chromosomes, Castrating Dwarves," 54.
173. The medical term for having an extra copy of a chromosome is trisomy. See CDC, "Down Syndrome."
174. Lombardo, "Tracking Chromosomes, Castrating Dwarves," 156.
175. Ibid., 154.
176. American Philosophical Society, Box 63: B: D27-2: Charles Davenport.
177. Ibid.
178. Kevles, *In the Name of Eugenics*, 80–81.
179. Ibid., 81.
180. Ibid.

181. Terms like *freak* and *monstrosity* were an acceptable part of the national vernacular at the time. Thus, usage of these terms is not meant to be disparaging. See Tanfer, "Freaks and Geeks," footnote 8.
182. Tanfer, "Freaks and Geeks," 3.
183. Ibid., 5.
184. Ibid., 6.
185. Davenport, "Medical Genetics and Eugenics," 14.
186. Tanfer, "Freaks and Geeks," 7.
187. Davenport, *State Laws Limiting Marriage Selection*, 34.
188. Tanfer, "Freaks and Geeks," 8.
189. Laughlin, "Preliminary Report," 14.

Chapter 8

190. Laughlin, *Eugenical Sterilization*, 325.
191. Ibid.
192. Laughlin, "Preliminary Report," 47.
193. *Brooklyn Daily Eagle*, February 26, 1914.
194. Black, *War Against the Weak*, 146.
195. Ibid., 145.
196. Ibid., 151.
197. Ibid., 153.
198. Ibid., 154
199. Ibid., 158.
200. Ibid., 162.
201. Ibid., 163.
202. Ibid., 164–65.
203. Ibid., 165.
204. Ibid., 167.
205. Ibid., 168.
206. Black, *War Against the Weak*, 169.
207. Black, at pages 172–173.

Chapter 9

208. Lombardo, *Three Generations, No Imbeciles*, 42.
209. Ibid.
210. Ibid., 47.
211. Lombardo, *Three Generations, No Imbeciles*, 42.
212. Ibid., 43.

213. Davenport, "Marriage Laws and Customs," 155.
214. *New York Times*, "Extends Work in Eugenics."
215. Lombardo, *Three Generations, No Imbeciles*, 47.
216. Ibid., 50.
217. Kaelber, "Compulsory Sterilization in 50 American States."
218. Kaelber, "Eugenics/Eugenic Sterilizations in Indiana."
219. Kaelber, "New Jersey."
220. Laughlin, *Eugenical Sterilization*, 142. Prior to January 1, 1922, the eugenic statutes in the following states were litigated in their state courts: Washington (1911–12), New Jersey (1912–13), Iowa (1914–17), Michigan (1916–18), New York (1915–20), Nevada (1915–18), Indiana (1919–21) and Oregon (1921–22).
221. Ibid., 25. The bill was introduced on March 5, 1912, by Assemblyman Robert P. Bush of Horseheads, New York. It passed the House on March 25, 1912, by a vote of 75–9 and the Senate on March 29, 1912, by a vote of 48–0 before being signed into law by Governor John A. Dix on April 16, 1912.
222. Laughlin, *Eugenical Sterilization*, 25.
223. Ibid., 86.
224. Ibid., 86.
225. Ibid., 87.
226. Ibid., 218.
227. Ibid., 218.
228. Ibid., 218–19.
229. The other board members were Charles H. Andrews, MD, and William J. Wansboro, MD.
230. Laughlin, *Eugenical Sterilization*, 222.
231. Ibid., 145. The New York sterilization law of 1912 was repealed on May 10, 1920. The bill to repeal the law was introduced by Senator Harry M. Sage of New York on April 8, 1920, and it unanimously passed by the Senate on April 12, 1920, 49–0. The repeal was signed by Governor Alfred E. Smith on May 10, 1920, and appears in the statutes of New York as "L. 1920, Chap. 619." See Laughlin, *Eugenical Sterilization*, 26.
232. A vasectomy is a minor surgical procedure that blocks sperm from reaching the semen that is ejaculated from the penis. See Urology Care Foundation, "What Is a Vasectomy?"
233. A salpingectomy is the surgical removal of one or more of the fallopian tubes. An ovariotomy, more commonly known as an oophorectomy, is the surgical removal of the ovaries.

234. Laughlin, *Eugenical Sterilization*, 144. At the time of this writing, no other data could be located to suggest that more sterilizations were legally performed pursuant to the New York sterilization statute, and given Laughlin's deep support for the law, there is nothing to suggest that he would have purposefully omitted any information about more procedures that may have been performed.
235. Ibid., 82.
236. Ibid., 96. The other states that had eugenic sterilization laws with the corresponding number of sterilizations in those states were Connecticut (27), Iowa (49), Kansas (54), Michigan (1), Nevada (0), New Jersey (0), New York (42), North Dakota (23), South Dakota (0) and Washington (1).

Chapter 10

237. Cohen, *Imbeciles*, 15.
238. Ibid., 16
239. Ibid., 78.
240. Ibid., 80.
241. Ibid., 82–83.
242. Lombardo, *Three Generations, No Imbeciles*, 106–7.
243. Ibid., 107.
244. Cohen, *Imbeciles*, 143.
245. Ibid., 144.
246. Ibid., 147.
247. Ibid., 149.
248. During Dr. Priddy's investigation, Priddy told Laughlin that there were several inmates at the colony named Buck and others named Harlowe, the latter of which was Emma's maiden name. Priddy believed that the Bucks and Harlowes, who both hailed from Albemarle County, were of the same family stock. While he could not prove it, Strode believed that Carrie was a Harlowe through her mother and that the "line of baneful heredity seems conclusive and unbroken." Cohen, *Imbeciles*, 147.
249. Cohen, *Imbeciles*, 149.
250. Ibid., 151.
251. Ibid., 152.
252. Ibid., 179.
253. Lombardo, *Three Generations, No Imbeciles*, 133.
254. Ibid., 148.

255. The initial proceedings were named *Buck v. Priddy*, with Dr. Priddy as the named actor on behalf of the colony. However, Dr. Priddy died on January 13, 1925, after the lower court ruled in favor of the colony. Dr. John Bell succeeded Priddy, and on appeal, the name of the case was changed to *Buck v. Bell*. Cohen, *Imbeciles*, 202.
256. Cohen, *Imbeciles*, 182–83.
257. Ibid., 184.
258. Ibid., 187.
259. Ibid.
260. Ibid., 188.
261. Ibid., 190.
262. Ibid., 192.
263. Ibid., 194.
264. Ibid.
265. Ibid., 201.
266. Ibid., 203.
267. Ibid., 208.
268. *Jacobson v. Massachusetts*, 197 U.S. 11 (1905).
269. Lombardo, *Three Generations, No Imbeciles*, 154.
270. *Buck v. Bell*, 274 U.S. 200 (1927). Associate Justice Pierce Butler was the lone dissenting judge in the case, although he did not issue a written opinion.
271. Ibid., 208.
272. Cohen, *Imbeciles*, 277.
273. This is by no means a defense of the overall decision by the Supreme Court in *Buck v. Bell*. The Court certainly had the authority to investigate any improprieties that may have existed in the lower court case, but it did not, and in many respects, the manner in which the case was brought to the highest court offered a convenient and predetermined decision, even if Justice Holmes wasn't a strong supporter of eugenics.
274. *Buck v. Bell*, 274 U.S. 200 (1927), 208.
275. Cohen, *Imbeciles*, 291.
276. Ibid., 286.
277. Ibid., 298.

Chapter 11

278. Laughlin, "Preliminary Report," 5.
279. Ibid., 55; Black, *War Against the Weak*, 58.
280. Ibid.

281. Ibid., 247.
282. Ibid., 249.
283. Ibid., 251.
284. Ibid., 253.
285. Ibid., *War Against the Weak*, 255.
286. Ibid., 259.
287. Wikipedia, "Hippocratic Oath."

Chapter 12

288. Black, *War Against the Weak*, 235.
289. Ibid., 236.
290. Ibid., 239.
291. Ibid., 240.
292. Ibid., 242.
293. Ibid., 243.
294. Ibid., 244.

Chapter 13

295. Ibid., 261.
296. Ibid., 264.
297. Ibid., 263.
298. Ibid., 264–65.
299. Ibid., 265.
300. Ibid., 267.
301. Ibid., 289.
302. Ibid., 292.
303. Ibid., 279.
304. Ibid., 282.
305. Ibid., 284.
306. Ibid., 286.
307. Ibid., 290.
308. Ibid., 294.
309. Ibid., 295.
310. Ibid., 271.
311. Ibid., 272.
312. Ibid., 276.
313. Ibid., 276.
314. *Eugenical News* 17, no. 2 (March/April 1932).

315. Black, *War Against the Weak*, 299.
316. Ibid., 304.
317. *Nassau (NY) Daily Review*, January 5, 1934.
318. Ibid., 315.
319. Ibid., 340.
320. Ibid., 342.
321. Ibid., 341
322. Ibid., 342.
323. Ibid., 344.
324. Ibid., 345.
325. Ibid., 346.
326. Ibid., 347.
327. Ibid., 348.
328. Ibid., 348.
329. Davenport, *Heredity*, 180.
330. Ibid., 350.
331. Ibid., 353.
332. Ibid., 355.
333. Ibid., 358.
334. Ibid., 360.
335. Ibid., 324.
336. Harry H. Laughlin Papers, Special Collections and Museums, Pickler Memorial Library, Truman State University, E-1-3, Hon. Degree.
337. Black, *War Against the Weak*, 312.
338. Remington, "Life Unworthy of Life," 2.
339. Ibid.
340. HLS Nuremberg Trials Project, "*U.S.A. v. Karl Brandt et al.*"
341. Remington, "Life Unworthy of Life," 17.
342. National Socialist. "Carl Schneider."

Chapter 14

343. Black, *War Against the Weak*, 388.
344. Ibid., 389.
345. Ibid., 413.
346. Ibid., 390.
347. Ibid.
348. Ibid.
349. Ibid., 393.
350. Ibid., 396.

351. Ibid., 384.
352. Riddle, *Biographical Memoir*, 91.
353. Ibid., 91.
354. Ibid., 387.
355. *Newsday*, March 25, 1944.
356. *New York Times*, March 9, 1946.

Chapter 15

357. Kramer, "Long-Range Studies," Table 1.
358. Gao, "Psychiatric Inpatient Beds."
359. Evana, "The Long Scalpel of the Law," Section 2.
360. Torpy, "Native American Women," 3–4.
361. Ibid.
362. Kaelber, "California."
363. *Los Angeles Daily News*, "LA County Seeks Reparations."
364. Tajima-Peña, *No Más Bebés*.
365. *Madrigal v. Quilligan*.
366. Ingram, "State Issues Apology."
367. Beam, "California Tries to Find."
368. Kaelber, "Eugenics/Sexual Sterilizations in North Carolina."
369. Shapiro, *State of Eugenics*.
370. Ibid.
371. Ibid.
372. Kaelber, "Virginia."
373. Portnoy, "Va. General Assembly Agrees."
374. Theobold, "1970 Law."
375. England, "Indian Health Service Policy," 2.
376. Ibid., 6–7.
377. Lawrence, "Indian Health Service," 407.
378. Torpy, "Native American Women," 1–22.
379. Library of Congress, "Puerto Rico
380. Denis, *War Against All Puerto Ricans*, 33–34.
381. At the time of this writing, no specific records depicting a direct connection between this program and officials at the Eugenics Record Office could be located. However, there is no doubt that the measures that were taken were directly inspired by or a consequence of the eugenics of the time, which, of course, were spearheaded by the ERO.
382. Back, Hill and Stycos, "Population Control in Puerto Rico," 563.

383. Womack, "U.S. Colonialism in Puerto Rico," 80.
384. Presser, "Puerto Rico," 343–61
385. Ibid., 345.
386. Cheng, "Some Dutch People."
387. National Women's Law Center, *Forced Sterilization*, 1.
388. In 1983, a New York court authorized the sterilization of a fourteen-year-old girl who was disabled. See *Application of Nilsson*, 122 Misc.2d 458 (1983). In 2002, another New York court followed the previous court's decision and also upheld a request to have a young woman sterilized. See *In re Guardianship of B*, 190 Misc.2d 581 (2002).
389. Sullivan, "Artificial Wombs for Premature Babies."
390. Firstpost, "Baby in a Pod."

Chapter 16

391. Carnegie Science, "Statement on Eugenics Research."
392. Trent, "World's Largest Body."
393. The American Society of Human Genetics was founded in 1948. The group currently boasts a membership of eight thousand and has a stated mission "to advance human genetics and genomics in science, health, and society through excellence in research, education, and advocacy." See American Association of Human Genetics, "About ASHG."
394. Ibid.

BIBLIOGRAPHY

Books

Black, Edwin. *War Against the Weak: Eugenics and America's Campaign to Create a Master Race.* Washington, D.C.: Dialog Press, 2003.

Briggs, Laura. *Reproducing Empire: Race, Sex, Science, and U.S. Imperialism in Puerto Rico.* Los Angeles: University of California Press, 2002.

Cohen, Adam. *Imbeciles: The Supreme Court, American Eugenics, and the Sterilization of Carrie Buck.* New York: Penguin, 2016.

Davenport, Charles Benedict. *Heredity in Relation to Eugenics.* New York: Henry Holt, 1911.

———. *Medical Genetics and Eugenics.* Philadelphia: Woman's Medical College of Pennsylvania, 1940.

———. *The Trait Book*. Cold Spring Harbor, New York, 1912.

Denis, Nelson A. *War Against All Puerto Ricans.* New York: Nation Books, 2015.

Goddard, Henry Herbert. *The Kallikak Family: A Study in the Heredity of Feeble-mindedness.* New York: MacMillan, 1912.

Kevles, Daniel J. *In the Name of Eugenics: Genetics and the Uses of Human Heredity.* London: First Harvard University Press, 1995.

Laughlin, Harry Hamilton. *Eugenical Sterilization in the United States.* Chicago: Psychopathic Laboratory of the Municipal Court of Chicago, 1922.

Lombardo, Paul A. *Three Generations, No Imbeciles: Eugenics, the Supreme Court and* Buck v. Bell. Baltimore, MD: Johns Hopkins Press, 2010.

Okrent, Daniel. *The Guarded Gate: Bigotry, Eugenics, and the Law That Kept Two Generations of Jews, Italians, and Other European Immigrants Out of America.* New York: Scribner, 2019.

Railey, John. *Rage to Redemption in the Sterilization Age: A Confrontation with American Genocide.* Eugene, OR: Cascade Books, 2015.

Sanger, Margaret. *Pivot of Civilization.* New York: Brentanos, 1922.

Schoen, Johanna. *Choice and Coercion: Birth Control, Sterilization, and Abortion in Public Health and Welfare.* Chapel Hill: North Carolina Press, 2005.

Stern, Alexandra Minna. *Eugenic Nation: Fault & Frontiers of Better Breeding in Modern America.* Los Angeles: University of California Press, 2005.

Articles and Special Collections

American Philosophical Institute. "Official Record of the Gift of the Eugenics Record Office, Cold Spring Harbor, Long Island, New York by Mrs. E.H. Harriman to the Carnegie Institution of Washington and of Its Acceptance by the Institution."

American Philosophical Society. AES, 57506: Am3, "Sermon #56: Religion and Eugenics AES Sermon Contest," 1927.

———. AES, Am3,575.06, "Proposed Itineraries of State Fairs," 1930.

———. B: D27-2, Box 126.

———. B: D27, Folder 5, Box 88.

———. Box 126: B: D27-2: Charles Davenport.

———. Box 62: B: D27-2: Charles Davenport.

———. Box 63: B: D27-2: Charles Davenport.

———. Box 63: B: D27-2.

———. Box Series I-63, Folder 1.

———. ERO, MSC77, Ser. X, Box 3: Harry M. Laughlin.

———. "Letter to Alexander Graham Bell." https://www.amphilsoc.org/item-detail/letter-alexander-graham-bell.

———. "Letter to Charles B. Davenport." https://www.amphilsoc.org/item-detail/letter-charles-b-davenport.

———. "Minutes of Meeting of Committee on Immigration of the Eugenics Research Association Held at the Harvard Club, New York City, at 11 a.m., February 25, 1920, APS Collections." https://www.amphilsoc.org/item-detail/committee-immigration.

Back, Kurt W., Rueben Hill and J. Mayone Stycos. "Population Control in Puerto Rico: The Formal and Informal Framework." *Law and Contemporary Problems* 25 (Summer 1960): 558–76. https://scholarship.law.duke.edu/lcp/vol25/iss3/12.

Bresnan, Emma. "Eugenics at William & Mary." William & Mary University. https://www.wm.edu/sites/lemonproject/_documents/bresnaneugenicsoutline.pdf.

Cohen, S. Adam. "Harvard's Eugenic Era." *Harvard Magazine* (March–April 2016). https://www.harvardmagazine.com/2016/02/harvards-eugenics-era.

Cold Spring Harbor. Box 9, Folders 9.1, 9.18, 9.25, 9.28, 9.29 and 9.32.

———. Box 14, Series 4 ERA Subseries ERO Memoirs, 1912.

———. Field Workers of the Eugenics Record Office, Tabulated Summary of Work (October 1, 1910, to January 1, 1913).

Davenport, Charles. "Marriage Laws and Customs." In *Problems in Eugenics: Papers Communicated to the First International Eugenics Congress* (London: Eugenics Education Society, 1912).

———. *State Laws Limiting Marriage Selection Examined in the Light of Eugenics.* Eugenics Record Office Bulletin No. 9. Cold Spring Harbor, NY: Eugenics Record Office, 1913.

Davenport, Charles, and Morris Steggerda. "The Shinnecock Indians of Long Island, New York." Unpublished manuscript, 1932. National Museum of Health and Medicine, Otis Historical Archives, No. 316, Box 71, File 8.

Dennert, James Walter. "Henry Herbert Goddard (1866–1957)." Embryo Project Encyclopedia, Arizona State University, May 6, 2021. https://embryo.asu.edu/pages/henry-herbert-goddard-1866-1957.

England, Charles R. "A Look at the Indian Health Service Policy in Sterilization 1972–1976." http://whale.to/b/england.html.

Eugenical News 17, no. 2 (March/April 1932). Cold Spring Harbor, Box 15 Subseries 3 Eugenical News 1916–1953, File 15.7.

"Eugenics in the Colleges." *Journal of Heredity* 5, no. 4 (April 1914):186. https://academic.oup.com/jhered/article-abstract/5/4/186/772008.

Evana, Brenna. "The Long Scalpel of the Law: How United States Prisons Continue to Practice Eugenics Through Forced Sterilization." *Minnesota Journal of Law and Inequality*, June 7, 2021. https://lawandinequality.org/2021/06/07/the-long-scalpel-of-the-law-how-united-states-prisons-continue-to-practice-eugenics-through-forced-sterilization.

Gao, Y. Nina. "The Relationship Between Psychiatric Inpatient Beds and Jail Populations in the United States." *Journal of Psychiatric Practice* 27, no. 1 (January 2021): 33–42. https://www.ncbi.nlm.nih.gov/pmc/articles/PMC7887772.

Gur-Arie, Rachel. "Harry Hamilton Laughlin (1880-1943)." Embryo Project Encyclopedia, Arizona State University, December 19, 2014. https://embryo.asu.edu/pages/harry-hamilton-laughlin-1880-1943.

Harry Laughlin Papers at the University of Missouri at Kirksville.

Harry H. Laughlin Papers, Special Collections and Museums, Pickler Memorial Library, Truman State University, E-1-3, Hon. Degree.

Hoff, Aliya R. "The International Eugenics Congresses (1912–1932)." Embryo Project Encyclopedia, July 29, 2021. https://hdl.handle.net/10776/13290.

Ingram, Carl. "State Issues Apology for Sterilization." *Los Angeles Times*, March 12, 2003. https://www.latimes.com/archives/la-xpm-2003-mar-12-me-sterile12-story.html.

Kaelber, Lutz. "California." University of Vermont. https://www.uvm.edu/~lkaelber/eugenics/CA/CA.html.

———. "Eugenics: Compulsory Sterilization in 50 American States." University of Vermont, March 4, 2009. https://www.uvm.edu/~lkaelber/eugenics.

———. "Eugenics/Eugenic Sterilizations in Indiana." University of Vermont. https://www.uvm.edu/~lkaelber/eugenics/IN/IN.html.

———. "Eugenics/Sexual Sterilizations in North Carolina." University of Vermont. https://www.uvm.edu/~lkaelber/eugenics/NC/NC.html.

———. "New Jersey." University of Vermont. https://www.uvm.edu/~lkaelber/eugenics/NJ/NJ.html.

———. "Virginia." University of Vermont. https://www.uvm.edu/~lkaelber/eugenics/VA/VA.html.

Kramer, Morton. "Long-Range Studies of Mental Hospital Patients: An Important Area for Research in Chronic Disease." *Milbank Memorial Fund Quarterly* 31, no. 3 (July 1953): 253–64. https://www.ncbi.nlm.nih.gov/pmc/articles/PMC2690271.

Laughlin, Harry. "Analysis of the Metal and Dross in America's Modern Melting Pot: The Determination of the Rate of Occurrence of the Several Definite Types of Social Inadequacy in Each of the Several Present Immigration and Native Population Groups in the United States." November 21, 1922. https://www.archive.org/details/analysisofameric00unit/page/n3/mode/2up.

Laughlin, Harry H. *Report of the Committee of the Eugenic Section of the American Horse Breeders' Association to Study and to Report on the Best Practical Means of Cutting Off the Defective Germ-Plasm in the Human Population*. Cold Spring Harbor, NY: Eugenics Record Office, 1914.

Lawrence, Jane. "The Indian Health Service and the Sterilization of Native American Women." *American Indian Quarterly* 24, no. 3 (Summer 2000): 400–419.

Lombardo, Paul A. "Medicine, Eugenics, and the Supreme Court: From Coercive Sterilization to Reproductive Freedom." *Journal of Contemporary Health Law & Policy* 13, no 1, article 5 (1996).

———. "Tracking Chromosomes, Castrating Dwarves: Uninformed Consent and Eugenic Research." Georgia State College of Law, Legal Studies Research Paper No. 2009-21. *Ethics and Medicine* 25, no. 149 (2009): 149–64.

National Women's Law Center. *Forced Sterilization of Disabled People in the United States.* 2021. https://nwlc.org/wp-content/uploads/2022/01/*f*. NWLC_SterilizationReport_2021.pdf.

Nursing Clio. "The Black Politics of Eugenics." June 1, 2017. https://nursingclio.org/2017/06/01/the-black-politics-of-eugenics.

Paul K. Longmore Institute on Disability. "Eugenics on Campus." https://longmoreinstitute.sfsu.edu/archive/eugenics-campus.html.

Presser, Harriett B. "Puerto Rico: The Role of Sterilization in Controlling Puerto Rican Fertility." *Population Studies* 23, no. 3 (November 1969): 343–61.

Remington, M. Alexander. "'Life Unworthy of Life' Aktion T4: The First Nazi Genocide." *Student Publication*s 1063 (Spring 2023). The Cupola: Scholarship at Gettysburg College.

Riddle, Oscar. *Biographical Memoir of Charles Benedict Davenport.* Washington, D.C.: National Academy of Sciences, 1947.

Schoen, Johanna. "Reassessing Eugenic Sterilization: The Case of North Carolina." *Rutgers University: Bioethics and Humanities* (2011): 141–60.

Singleton, Marilyn M. "The Science of Eugenics: America's Moral Detour." *Journal of American Physicians and Surgeons* 19, no. 4 (Winter 2014).

Strong, John A. "Miss-measuring the Unkechaug: The Reservation as a Eugenics Laboratory, 1923." Unpublished report presented at the Ethnohistory Conference, Missouri State University, 2012.

———. Presentation. "Legacy of Long Island's First Peoples, Program 2-Eugenics Study of Long Island's Native Americans." https://www.waltwhitman.org/events/eugenics-study-of-long-islands-native-americans.

Tanfer, Emin-Tunc. "Freaks and Geeks: Coney Island Sideshow Performers and Long Island Eugenicists, 1910–1935." *Long Island Historical Journal* 14, nos. 1–2 (Fall 2001–Spring 2002): 1–14.

Torpy, Sally J. "Native American Women and Coerced Sterilization: On the Trail of Tears in the 1970s." *American Indian Culture and Research Journal* 24, no. 2 (2000): 1–22.

University of Missouri. "Harry Laughlin: Workhorse of the American Eugenics Movement." Curated 2011. https://library.missouri.edu/specialcollections/exhibits/show/controlling-heredity/america/laughlin.

Womack, Malia Lee. "U.S. Colonialism in Puerto Rico: Why Intersectionality Must Be Addressed in Reproductive Rights." *St. Anthony's International Review* 16, no. 1 (2020): 74–85.

Legal Cases

Application of Nilsson, 122 Misc.2d 458 (1983).
Buck v. Bell, 274 U.S. 200 (1927).
In re Guardianship of B, 190 Misc.2d 581 (2002).
Jacobson v. Massachusetts, 197 U.S. 11 (1905).
Madrigal v. Quilligan, United States District Court for the Central District of California, No. CV 75-2057-JWC (1978).
New York v. Osborn, 103 Misc. Rep. 23, 169 N. Y. Sup. 638, 171 N. Y. Sup. 1094 (1915).

Newspaper Articles

Beam, Adam. "California Tries to Find and Compensate Victims of Forced Sterilization Program." Associated Press, January 7, 2023.
Brooklyn Daily Eagle. "For Biologists: A Charming Retreat at Cold Spring Harbor." July 8, 1890.
———. "Inheritance Atom Proves a Complex Unit in Evolution." June 28, 1929.
———. "Meeting of Eugenicists." September 29, 1921.
———. "Scientists at Cold Spring Harbor Seek to Unravel Life's Riddle." December 25, 1910.
———. Untitled. February 26, 1914.
———. Untitled. July 14, 1930.
———. Untitled. March 29, 1914.
Cheng, Maria. "Some Dutch People Seeking Euthanasia Cite Autism or Intellectual Disabilities, Researchers Say." NBC Philadelphia, updated June 28, 2023. https://www.nbcphiladelphia.com/news/national-international/some-dutch-people-seeking-euthanasia-cite-autism-or-intellectual-disabilities-researchers-say/3594172.
Corrector (Sag Harbor, NY). June 6, 1874.
County Review (Riverhead, NY). "Harris-Davenport," August 25, 1922.
East Hampton (NY) Star, June 7, 1912; June 27, 1913.
Glen Falls (NY) Times, August 9, 1933.
Long-Island (NY) Star, May 26, 1819.
Long-Islander (Huntington, NY). "Cold Spring." April 23, 1898.
———. "Cold Spring." June 9, 1911.
———. "Rockefeller Aids Work." April 4, 1913.
———. "Want Town Board to Make Changes." November 8, 1929.
Los Angeles Daily News. "LA County Seeks Reparations for 1968–1974 Forced Sterilizations at LAC+USC Med Center." July 13, 2021.

Nassau (NY) Daily Review, January 5, 1934.
Newsday, March 25, 1944.
New York Times. "Extends Work in Eugenics," March 30, 1913.
———. "Mrs. E.H. Harriman Dies at the Age of 81." November 8, 1932
———. Untitled. March 9, 1946.
Northport (NY) Observer, November 1, 1929; January 10, 1930.
Portnoy, Jenna. "Va. General Assembly Agrees to Compensate Eugenics Victims." *Washington Post*, February 27, 2015.
Trent, Sydney. "World's Largest Body of Human Geneticists Apologizes for Eugenics Role." *Washington Post*, January 23, 2023.

Documentaries and Films

Browning, Tod. *Freaks.* 1932.
Ferrari, Michelle. *The Eugenics Crusade.* American Experience, PBS, October 16, 2018.
Shapiro, Dawn Sinclair. *The State of Eugenics*. 2016.
Tajima-Peña, Renee. *No Más Bebés.* Independent Lens, June 14, 2015.
Wharton, Leopold, and Theodore Wharton. *The Black Stork.* 1917.

Websites

American Association of Human Genetics. "About ASHG." https://www.ashg.org/about.
Archbold, Matthew. "7 Beloved Famous People Who Were Wildly Pro-Eugenics." *National Catholic Register*, November 14, 2014. https://www.ncregister.com/blog/7-beloved-famous-people-who-were-wildly-pro-eugenics.
Asylum Postcards. www.asylumpostcards.com.
Brittanica. "Scientific Method." Last updated July 31, 2024. https://www.britannica.com/science/scientific-method.
Carnegie Science. "Statement on Eugenics Research from Carnegie President Eric D. Isaacs." August 12, 2020. https://carnegiescience.edu/about/history/statement-eugenics-research.
CDC. "Down Syndrome." May 16, 2024. https://www.cdc.gov/birth-defects/about/down-syndrome.html.
Cohen, Lon. "Historical Facts About Kings Park Psychiatric Center." LongIsland.com, May 24, 2022. https://www.longisland.com/news/05-19-22/historical-facts-about-kings-park-psychiatric-center.html.
Cold Spring Harbor Laboratory. "Charles B. Davenport." https://www.cshl.edu/personal-collections/charles-b-davenport.

———. "Eugenics Record Office." https://www.cshl.edu/archives/institutional-collections/eugenics-record-office.

———. "Nam Family Members, Photo by Arthur Estabrook, 1912." https://dnalc.cshl.edu/view/15724-Nam-family-members-photo-by-Arthur-Estabrook-1912.html

Firstpost. "Baby in a Pod: What Is EctoLife, the World's 'First Artificial Womb Facility'?" December 14, 2022. https://www.firstpost.com/explainers/ectolife-the-worlds-first-artificial-womb-facility-11805801.html.

HLS Nuremberg Trials Project. "NMT Case 1: *U.S.A. v. Karl Brandt et al.*: The Doctors' Trial." Last reviewed March 2020. https://nuremberg.law.harvard.edu/nmt_1_intro.

Levine, David. "The Real History of Letchworth Village in the Hudson Valley." Hudson Valley, January 18, 2020. https://www.hvmag.com/life-style/letchworth-village-thiells.

Library of Congress. "Puerto Rico and the United States." https://www.loc.gov/collections/puerto-rico-books-and-pamphlets/articles-and-essays/nineteenth-century-puerto-rico/puerto-rico-and-united-states.

Loving-Long-Island.com. "Cold Spring Harbor, Long Island, NY." http://www.loving-long-island.com/cold-spring-harbor.html.

National Park Service. "The Panama-Pacific International Exhibition." Last updated July 30, 2024. https://www.nps.gov/goga/learn/historyculture/ppie.htm.

National Socialist. "Carl Schneider (1891–1946)." https://www.t4-denkmal.de/eng/Carl-Schneider.

Social Networks and Archival Context. "Brooklyn Institute of Arts and Sciences." https://snaccooperative.org/ark:/99166/w6kq1vrx.

Sullivan, Will. "Artificial Wombs for Premature Babies Might Soon Begin Human Trials." *Smithsonian Magazine*, September 26, 2023. https://www.smithsonianmag.com/smart-news/artificial-wombs-for-premature-babies-might-soon-begin-human-trials-180982949.

Theobold, Brianna. "A 1970 Law Led to the Mass Sterilization of Native American Women. That History Still Matters." *TIME*, last updated November 28, 2019. https://time.com/5737080/native-american-sterilization-history.

Urology Care Foundation. "What Is a Vasectomy?" Last updated December 2020. https://www.urologyhealth.org/urology-a-z/v/vasectomy.

Vassar Encyclopedia. "Dr. Katharine Bement Davis 1892." https://www.vcencyclopedia.vassar.edu/distinguished-alumni/katharine-bement-davis.

Wikipedia. "Hippocratic Oath." https://en.wikipedia.org/wiki/Hippocratic_Oath.

———. "Madison Grant." https://en.wikipedia.org/wiki/Madison_Grant.

ABOUT THE AUTHOR

Mark A. Torres is the author of *Long Island Migrant Labor Camps: Dust for Blood* (2021); two crime novels, *A Stirring in the North Fork* (2015) and *Adeline* (2019); and a labor union–related children's book titled *Good Guy Jake* (2017). Mark is also a labor and employment attorney who tirelessly represents thousands of unionized workers and their families throughout the Greater New York area. His commitment to the labor movement spans more than thirty years. Mark has a law degree from Fordham University School of Law and a bachelor's degree in history from New York University. He is currently an adjunct professor of labor studies at Hofstra University.